**Books are to be returned on or before
the last date below.**

Taints and off-flavours in food

Related titles from Woodhead's food science, technology and nutrition list:

Texture in semi-solid foods (ISBN 1 85573 673 X)

Understanding and controlling the texture of semi-solid foods such as yoghurt and ice cream is a complex process. With a distinguished international team of contributors, this important collection summarises some of the most important research in this area. The first part of the book looks at the behaviour of gels and emulsions, how they can be measured and their textural properties improved. The second part of the collection discusses the control of texture in particular foods such as yoghurt, ice cream, spreads and sauces.

Colour in food (ISBN 1 85573 590 3)

The colour of a food is central to consumer perceptions of quality. This important new collection reviews key issues in controlling colour quality in food, from the chemistry of colour in food to measurement issues, improving natural colour and the use of colourings to improve colour quality.

The nutrition handbook for food processors (ISBN 1 85573 464 8)

Over the past decade there has been an explosion of research on the relationship between nutrition and health. Given the current interest of consumers in more nutritious food, food processors are increasingly concerned with understanding the nature of nutrient loss during food processing. This new book brings together an international team of experts to summarise key findings on diet and nutrient intake, the impact of nutrients on health, and how food processing operations affect the nutritional quality of foods.

Details of these books and a complete list of Woodhead's food science, technology and nutrition titles can be obtained by:

- visiting our web site at www.woodhead-publishing.com
- contacting Customer Services (e-mail: sales@woodhead-publishing.com; fax: +44 (0) 1223 893694; tel.: +44 (0) 1223 891358 ext. 30; address: Woodhead Publishing Ltd, Abington Hall, Abington, Cambridge CB1 6AH, England)

Selected food science and technology titles are also available in electronic form. Visit our web site (www.woodhead-publishing.com) to find out more.

If you would like to receive information on forthcoming titles in this area, please send your address details to: Francis Dodds (address, tel. and fax as above; e-mail: francisd@woodhead-publishing.com). Please confirm which subject areas you are interested in.

Taints and off-flavours in food

**Edited by
Brian Baigrie**

CRC Press
Boca Raton Boston New York Washington, DC

WOODHEAD PUBLISHING LIMITED
Cambridge, England

Published by Woodhead Publishing Limited, Abington Hall, Abington
Cambridge CB1 6AH, England
www.woodhead-publishing.com

Published in North America by CRC Press LLC, 2000 Corporate Blvd, NW
Boca Raton FL 33431, USA

First published 2003, Woodhead Publishing Ltd and CRC Press LLC
© 2003, Woodhead Published Ltd
The authors have asserted their moral rights.

This book contains information obtained from authentic and highly regarded sources. Reprinted material is quoted with permission, and sources are indicated. Reasonable efforts have been made to publish reliable data and information, but the authors and the publishers cannot assume responsibility for the validity of all materials. Neither the authors nor the publishers, nor anyone else associated with this publication, shall be liable for any loss, damage or liability directly or indirectly caused or alleged to be caused by this book.
  Neither this book nor any part may be reproduced or transmitted in any form or by any means, electronic or mechanical, including photocopying, microfilming and recording, or by any information storage or retrieval system, without permission in writing from the publishers.
  The consent of Woodhead Publishing and CRC Press does not extend to copying for general distribution, for promotion, for creating new works, or for resale. Specific permission must be obtained in writing from Woodhead Publishing or CRC Press for such copying.

Trademark notice: Product or corporate names may be trademarks or registered trademarks, and are used only for identification and explanation, without intent to infringe.

British Library Cataloguing in Publication Data
A catalogue record for this book is available from the British Library.

Library of Congress Cataloging in Publication Data
A catalog record for this book is available from the Library of Congress.

Woodhead Publishing ISBN 1 85573 449 4 (book) 1 85573 697 7 (e-book)
CRC Press ISBN 0-8493-1744-4
CRC Press order number: WP1744

Typeset by SNP Best-set Typesetter Ltd., Hong Kong
Printed by TJ International, Padstow, Cornwall, England

Contents

List of contributors... viii

1 Introduction... 1
B. Baigrie, Reading Scientific Services Ltd, UK
   1.1   Defining off-flavours and taints....................... 1
   1.2   The structure of this book............................ 3
   1.3   References.. 4

**2 Sensory analytical methods in detecting taints
and off-flavours in food**................................. 5
D. Kilcast, Leatherhead Food International, UK
   2.1   Introduction.. 5
   2.2   Sensory test procedures.............................. 10
   2.3   Types of test: difference/discrimination tests........ 14
   2.4   Types of test: quantitative and hedonic tests......... 16
   2.5   Data handling, analysis and presentation.............. 20
   2.6   Choosing and interpreting sensory tests............... 20
   2.7   Applications of sensory testing....................... 23
   2.8   Standardisation of test methods and instrumental methods.. 26
   2.9   Ethical aspects....................................... 28
   2.10 Future trends... 28
   2.11 References.. 29

3 Instrumental methods in detecting taints and off-flavours......... 31
W. J. Reid, Leatherhead Food International, UK
   3.1   Introduction.. 31

3.2	Liquid-based extraction techniques	32
3.3	Headspace extraction	37
3.4	Solid phase microextraction (SPME)	40
3.5	Gas chromatography and other methods	42
3.6	Stir-bar sorptive extraction	47
3.7	Electronic noses	48
3.8	References	54

4 Packaging materials as a source of taints 64
T. Lord, Pira International, UK

4.1	Introduction	64
4.2	Main types of food packaging	66
4.3	Sources of taints	70
4.4	Chemicals responsible for taints	75
4.5	Main foodstuffs affected	78
4.6	Instrumental analysis of taints	80
4.7	Sample preparation techniques	81
4.8	Sampling strategy	88
4.9	Examples of taint investigations	88
4.10	Preventing taints	101
4.11	Developments in taint monitoring: electronic noses	104
4.12	Tracing the cause of a packaging taint	106
4.13	Future packaging trends affecting taints	107
4.14	Sources of further information and advice	108
4.15	References	109

5 Microbiologically derived off-flavours 112
F. B. Whitfield, Food Science Australia, Australia

5.1	Introduction	112
5.2	Bacteria	113
5.3	Aerobic bacteria	117
5.4	Facultative anaerobic bacteria	125
5.5	Anaerobic bacteria	128
5.6	Actinomycetes	129
5.7	Fungi	130
5.8	Future trends	135
5.9	References	136

6 Oxidative rancidity as a source of off-flavours 140
R. J. Hamilton, formerly of Liverpool John Moores University, UK

6.1	Introduction	140
6.2	Oxidation	142
6.3	Autoxidation	143
6.4	Photo-oxidation	146
6.5	Lipoxygenase (LOX)	149

Contents   vii

    6.6   Ketonic rancidity and metal-catalysed lipid oxidation 151
    6.7   Off-flavours from volatile lipid molecules 152
    6.8   Case study: lipid autoxidation and meat flavour
           deterioration. 153
    6.9   Case study: lipid oxidation in fish. 156
    6.10   Conclusions: preventing off-flavours 157
    6.11   Sources of further information and advice. 158
    6.12   References 158

7   The Maillard reaction as a source of off-flavours 162
    A. Arnoldi, University of Milan, Italy
    7.1   Introduction 162
    7.2   Mechanism of the Maillard reaction 163
    7.3   Relevant Maillard reaction products (MRPs) in food
           flavour 166
    7.4   Food staling and off-flavours in particular foods 170
    7.5   References 173

8   Off-flavours due to interactions between food components 176
    E. Spinnler, INRA, France
    8.1   Introduction 176
    8.2   Flavour compound volatility in different food matrices 176
    8.3   Flavour retention in different food matrices 180
    8.4   Off-flavours caused by reactions between components
           in the food matrix 181
    8.5   Bacterial interactions with the food matrix causing
           off-flavours 182
    8.6   Bacterial interactions with additives causing off-flavours ... 183
    8.7   Conclusion: identifying and preventing off-flavours. 184
    8.8   References 186

9   Taints from cleaning and disinfecting agents 189
    C. Olieman, NIZO food research, The Netherlands
    9.1   Introduction 189
    9.2   Cleaning and disinfecting agents 190
    9.3   Testing the safety of cleaning and disinfecting agents 192
    9.4   Testing cleaning and disinfecting agents for their
           capacity to cause taints. 193
    9.5   Detecting cleaning and disinfecting agents in rinse water ... 193
    9.6   Detecting cleaning and disinfecting agents in food 194
    9.7   Measurement of active chlorine residues via chloroform ... 195
    9.8   Future trends 197
    9.9   References 197

Index 199

Contributors

Chapter 1

B. Baigrie
Reading Scientific Services Limited
Lord Zuckerman Research Centre
Whiteknights
PO Box 234
Reading
RG6 6LA
UK

Tel: +44 (0) 118 9868541
Fax: +44 (0) 118 9868932
E-mail: brian.d.baigrie@rssl.co.uk

Chapter 2

D. Kilcast
Leatherhead Food International
Randalls Road
Leatherhead
Surrey
KT22 7RY
UK

Tel: +44 (0) 1372 822321
Fax: +44 (0) 1372 836228
E-mail: dkilcast@leatherheadfood.com

Chapter 3

W. J. Reid
Leatherhead Food International
Randalls Road
Leatherhead
Surrey
KT22 7RY
UK

Tel: +44 (0) 1372 822213
Fax: +44 (0) 1372 836228
E-mail: breid@leatherheadfood.com

Chapter 4

T. Lord
Pira International
Randalls Road
Leatherhead
Surrey
KT22 7RU
UK

Tel: +44 (0) 1372 802000
Fax: +44 (0) 1372 802238
E-mail: tlord@pira.co.uk

Chapter 5

F. B. Whitfield
Food Science Australia
16 Julius Avenue
Riverside Corporate Park
Delhi Road (PO Box 52)
North Ryde
NSW 1670
Australia

Tel: +61 2 9490 8333
Fax: +61 2 9490 8499
E-mail: frank.whitfield@foodscience.afisc.csiro.au

Chapter 6

R. J. Hamilton
Formerly of Liverpool John Moores University
10 Norris Way
Formby
L37 8DB
UK

Tel: +44 (0) 1704 877572
Fax: +44 (0) 1704 877572

Chapter 7

A. Arnoldi
DISMA
University of Milan
Via Celoria 2
20133 Milano
Italy

Tel: +39 02 503 16806
Fax: +39 02 503 16801
E-mail: arnoldi@mailserver.unimi.it

Chapter 8

E. Spinnler
GER de Technologies et Procedes Alimentaires
CBAI INA-PG
78850 Thiverval
France

Tel: 01 30 81 53 87
Fax: 01 30 81 55 97
E-mail: spinnler@grignon.inra.fr

Chapter 9

C. Olieman
NIZO food research
2 Kernhemseweg
PO Box 20
6710 BA Ede
The Netherlands

Tel: +31 318 659 511
Fax: +31 318 650 400
E-mail: kees.olieman@nizo.nl

1
Introduction

B. Baigrie, Reading Scientific Services Ltd, UK

1.1 Defining off-flavours and taints

Flavour is one of the most important sensory qualities of a food. Whilst the appearance and colour of a food provide the first indicator of quality, its flavour and texture are critical in confirming or undermining that initial impression (Cardello, 1994). The flavour of a food is determined by a complex mix of taste, aroma, chemical response and texture (Meilgaard et al., 1999). Taste is the response of the taste buds on the tongue to particular soluble, involatile materials which trigger a range of basic taste sensations (variously categorised as sweet, sour, salty, bitter and umami). Aroma is the response of the olfactory epithelium in the roof of the nasal cavity to volatiles entering the nasal passage. The trigeminal nerve also responds to certain chemical irritants which may give a heating or cooling response. Finally, the texture of a food and the way it breaks down in the mouth determine how tastants, odorants and trigeminal stimulants are released. Moreover, the appearance and colour of the food, the physical perception of its texture and the sounds produced by the chewing of the food, will also affect how flavour is perceived and interpreted (Lawless and Heymann, 1998). Finally, factors such as genetic differences and individual experience also play a role in what we perceive and how we respond to that perception.

Off-flavours are widely defined as unpleasant odours or tastes resulting from the natural deterioration of a food (British Standards Institution, 1992). Taints are generally regarded as unpleasant odours or tastes resulting from contamination of a food by some foreign chemical with which it accidentally comes into contact. Useful definitions for the purposes of this book are:

2   Taints and off-flavours in food

- **Taint:** an atypical odour or taste resulting from internal deterioration in the food
- **Off-flavour:** an atypical odour or taste caused by contamination by a chemical foreign to the food.

Off-flavours and taints can be triggered by the involatiles detected by taste, but the main chemicals involved are the volatiles involved in odour response. Whilst there is a relatively small range of involatile tastants and trigeminal stimulants, over 200 individual odour qualities have been identified. Odours and aromas are usually composed of hundreds of compounds carried in an air stream which itself contains many other compounds. Since there are an estimated 1000 olfactory receptor genes influencing an individual's perception of odours, making it possible for the human nose to detect very low concentrations of some of these compounds (below 0.01 parts per billion), identifying an off-flavour or taint can be a particularly challenging process.

Off-flavours and taints can occur in many types of food for a variety of reasons. Table 1.1 illustrates the nature of the problem by showing, on the one hand, examples of microbiologically derived off-flavours and, on the

Table 1.1   Examples of sources of taints and off-flavours (source: Reading Scientific Services Ltd)

Source of taint/off-flavour	Compounds involved	Food contaminated
Microbiological	Indoles, skatole	Cereals
	Guaiacol	Custard
	Guaiacol	Fruit juice
	Paracresol	Milk
	Ethyl acetate, ethanol	Orange juice
Packaging:		
—printed cartonboard	Aliphatic hydrocarbons	Chocolates
—printed cartonboard	2-Butoxyethanol	Ice cream
—paper board	Alkoxy alcohols	Milk
—can laquer	Orthocresol	Cola
—paper sacks	Chloroanisoles	Desiccated coconut
—sealing foil	Methyl acrylic acid	Peanut butter
Other sources:		
—storage in proximity to fuel and solvent spills	Hydrocarbons	Carbonated beverages in PET bottles
—proprietary cleaner	Chlorophenols	Cheese
—contaminated process water	Methyl isoborneol	Distilled spirit
—contaminated tank	6-Orthochlorocresol	Jam filling
—cardboard boxes used to store empty bottles	Chloroanisoles	Spring water
—factory flooring	Xylenols	Chocolate eggs

Table 1.2 Foodstuffs affected by taints (source: PIRA)

Foodstuff	%
Sweets	30
Cakes and biscuits	20
Beverages	15
Bread/pizza	10
Soups	5
Alcoholic drinks	5
Cheese	5
Crisps	5
Other	5

other, taints caused by a variety of contaminants. Table 1.2 shows some of the foodstuffs most commonly affected by taints. Whatever the cause, the occurrence of a taint and off-flavour problem can prove to be extremely costly for a food manufacturer. In addition to the direct costs of a product recall, lost revenue and production downtime, there may be even more serious consequences in terms of litigation, adverse publicity, lower sales and a damaged brand image. Understanding the nature and causes of taints and off-flavours, and developing appropriate means of identifying them so that they can be either prevented or identified rapidly, remain major preoccupations of the food industry. These are also the themes of this book.

1.2 The structure of this book

Since the processes involved in flavour perception are so complex, the first stage in the analysis of a taint is usually sensory evaluation of the affected product using a panel of experts trained in detecting and describing taints. The ability of panellists to describe accurately the sensory properties of taints and to relate them to known standards, is particularly useful in providing information about the chemical nature of the taint and in suggesting avenues for subsequent instrumental analysis. Chapter 2 reviews sensory methods in detecting and analysing taints and off-flavours in food, looking, in particular, at the various types of test available, and how to choose between them and interpret the results. After sensory evaluation, a concentrated flavour extract is prepared from the tainted product and then fractionated into individual compounds which can be analysed and identified. Chapter 3 looks at the range of extraction techniques and the use of gas chromatography to fractionate and identify tainting compounds. It also considers new instrumental techniques such as stir-bar sorptive extraction and, in particular, the development of electronic noses.

4   Taints and off-flavours in food

The following group of chapters then look at some of the main causes of taints and off-flavours. Chapter 4 reviews the most important single cause of taints in food: packaging. It also builds on the previous chapters by looking in detail at how packaging taints are identified in practice as well as how they can be prevented. Chapter 5 discusses microbiologically derived off-flavours and the range of bacteria, fungi and yeasts responsible, the foods they affect and the particular off-flavours produced. Together with microbiological spoilage, oxidative rancidity is one of the principal causes of off-flavours and is discussed in Chapter 6. Chapter 7 then reviews another important process: the influence of the Maillard reaction.

1.3   References

BRITISH STANDARDS INSTITUTION (1992), BS 5098 (ISO 6658): 'Glossary of terms relating to sensory analysis'.

CARDELLO A V (1994), 'Consumer expectations and their role in food acceptance', in MacFie H and Thomson D (eds), *Measurement of Food Preferences*, Blackie Academic and Professional, London.

LAWLESS H and HEYMANN H (1998), *Sensory Evaluation of Food: Principles and Practice*, Chapman and Hall, London.

MEILGAARD M, CIVILLE G and CARR B (1999), *Sensory Evaluation Techniques*, 3rd edn, CRC Press, Boca Raton.

2
Sensory analytical methods in detecting taints and off-flavours in food

D. Kilcast, Leatherhead Food International, UK

2.1 Introduction

In the increasingly affluent societies of the developed countries, consumers expect a wide choice of food that is safe, enjoyable to eat, nutritious and of consistent quality. If food falls short of these criteria, there is sufficient choice in a commercially competitive environment for consumers to change their allegiances and find alternative products. Provided that their food is safe, consumers will view taste or flavour as the most important quality attribute. In particular, they will not tolerate flavours, commonly called taints or off-flavours, that are foreign to that product.

In common usage, the word 'taint' inevitably implies something that is unpleasant and undesirable. Within the context of food, definitions become more precise. The definition of taint (BSI, 1992) accepted in Standards is a taste or odour foreign to the product. This Standard also distinguishes an off-flavour as an atypical flavour usually associated with deterioration. These definitions are characterised by an important distinction from dictionary definitions: food taints are perceived by the human senses. This does not diminish the undesirable nature of chemical contamination of foods, but focuses on those contaminants that can be perceived, particularly by their odour or flavour, and which can be perceived at extremely low concentrations, for example parts per million (ppm), 10^6, parts per billion (ppb), 10^9, or even parts per trillion (ppt), 10^{12}. One important aspect of the definitions is that they do not incorporate any aspect of toxicity, whereas in common usage 'taint' is often associated with a risk to health. This may be one of the reasons why taint problems can be emotive in a consumer context. For practical reasons, more specific definitions can be proposed:

6   Taints and off-flavours in food

- **Taints:** unpleasant odours or flavours imparted to food through external sources.
- **Off-flavours:** unpleasant odours or flavours imparted to food through internal deteriorative change.

Although the two types of taint can render food equally unpleasant, this distinction is of great assistance in identifying the cause of taint problems. The difficulties in identifying taint problems result from a number of sources. First, consumer descriptions of taint are, with a few exceptions, notoriously unreliable, partly because of a lack of any training in analytical descriptive methods but mainly because of unfamiliarity with the chemical species responsible for taint. One notable exception is taint resulting from chlorophenol contamination, which is reliably described as antiseptic, 'TCP' or medicinal, this reliability being a consequence of consumer familiarity with products characterised by these sensations. Second, the extremely low concentrations that can give rise to taint present immense difficulties for the analyst who tries to identify the chemical nature of the taint. Third, taint can occur at all stages of the food manufacture and supply chain, and from many different sources at each stage. Consequently, the detective work needed to identify the cause of taint-oriented consumer complaints can be quite different for taints than for off-flavours.

The highly unpleasant nature of food taints can generate severe problems for retailer, producer, ingredients supplier, farmer, equipment supplier and even building contractor. These problems can include lost production, lost sales, lost consumer confidence, damaged brand image, damaged commercial relationships between supplier, manufacturer and retailer, and expensive litigation proceedings. Food manufacturers readily understand the financial implications of a day's defective production, but not the more widespread implication of tainted production; for example, the tainted production will not be reworkable and the plant may suffer from extended shut-down whilst defective building materials are replaced. In addition, identifying the source of a taint can be time consuming and expensive.

The principal difficulty in protecting against taint is the extremely low levels of contamination that can give rise to unpleasant characteristics. Examples of the low concentrations at which the most commonly encountered tainting materials (halophenols and haloanisoles) can be perceived are shown in Table 2.1.

2.1.1   The human senses

In order to understand how these low levels can be detected, the nature of the human perceptual processes must be considered. Human beings employ a range of senses in perceiving food quality (Table 2.2). The discussion below summarises these senses briefly. Fuller descriptions can be found in

Table 2.1 Taste thresholds (in water) of halophenols and haloanisoles

Compound	Parts per billion (10^9)
2-Chlorophenol	0.1
2-Bromophenol	0.03
2,6-Dichlorophenol	0.3
2,6-Dichloroanisole	0.04 (odour)
2,6-Dibromophenol	5×10^{-4}
2,4,6-Trichlorophenol	2
2,4,6-Trichloroanisole	0.02
2,4,6-Tribromophenol	0.6
2,4,6-Tribromoanisole	8×10^{-6} (odour)

Table 2.2 The human senses

Sense	Perception	
Vision	Appearance	
Gustation	Taste	
Olfaction	Odour/aroma	Flavour
Chemical/trigeminal	Irritant	
Touch	Texture	
Hearing	Texture	

the standard sensory texts, for example Hutchings (1994), Lawless and Heymann (1998), Meilgaard *et al.* (1999), Piggott (1988) and Rosenthal (1999).

The visual senses are of particular importance in generating an initial impression of food quality that often precedes the input from the remaining senses. If the appearance of the food creates a negative impact, then the other senses might not come into play at all. The visual sense is often related only to colour, but provides input on many more appearance attributes that can influence food choice. In particular, the visual senses can provide an early and strong expectation of the flavour and textural properties of foods.

Taste (gustation) is strictly defined as the response by the tongue to soluble, involatile materials. These have classically been defined as four primary basic taste sensations: salt, sweet, sour and bitter, although in some countries this list is extended to include sensations such as metallic, astringency and umami, this last sensation being associated with monosodium glutamate. The taste receptors are organised groups of cells, known as taste buds, located within specialised structures called papillae. These are located mainly on the tip, sides and rear upper surface of the tongue. Sweetness is detected primarily on the tip of the tongue, salt and sour on the sides of the tongue, and bitter on the rear of the tongue. Taste stimuli are characterised by the relatively narrow range between the weakest and the strongest stimulants (ca. 10^4) and are strongly influenced by factors such as temperature and pH.

The odour response is much more complex. Odours are detected as volatiles entering the nasal passage, either directly via the nose or indirectly through the retronasal path via the mouth. The odorants are sensed by the olfactory epithelium, which is located in the roof of the nasal cavity. Some 150–200 odour qualities have been recognised. There is a very wide range (ca. 10^{12}) between the weakest and the strongest stimulants. The odour receptors are easily saturated and specific anosmia (blindness to specific odours) is common. It is thought that the wide range of possible odour responses contributes to variety in flavour perception. Both taste and odour stimuli can be detected only if they are released effectively from the food matrix during the course of mastication.

The chemical sense corresponds to a pain response through stimulation of the trigeminal nerve. This is produced by chemical irritants such as ginger and capsaicin (from chilli), both of which give a heat response, and chemicals such as menthol and sorbitol, which give a cooling response. With the exception of capsaicin, these stimulants are characterised by high thresholds. The combined effect of the taste, odour and chemical responses gives rise to the sensation generally perceived as flavour, although these terms are often used loosely.

Texture is perceived by the sense of touch and comprises two components: somesthesis, a tactile, surface response from skin, and kinesthesis (or proprioception), which is a deep response from muscles and tendons For many foods, visual stimuli will generate an expectation of textural properties. The touch stimuli themselves can arise from tactile manipulation of the food with the hands and fingers, either directly or through the intermediary of utensils such as a knife or spoon. Oral contact with food can occur through the lips, tongue, palate and teeth, all of which provide textural information (Meilgaard et al., 1999). The texture of products such as snack foods and hard fruits is also influenced by sounds emitted during mastication.

The descriptions given above, whilst appropriate for the individual sensing modalities, fail to take into account their interactive nature, shown schematically in Fig. 2.1. Visual appearance factors strongly influence the expectation of flavour, but the most important interaction in the perception of taint is that between texture and flavour. The importance of this lies in the physical processes, in particular structural and textural changes, that occur during mastication. As food enters the mouth and is either bitten or manipulated between tongue and palate, catastrophic changes occur to the structure of the food which strongly influence the way in which tastants,

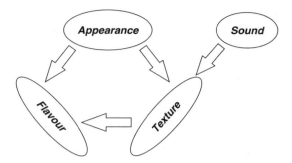

Fig. 2.1 The human senses and their interactions.

odorants and trigeminal stimulants are released from the food. The chemical species that give rise to taint can fall into each of these three classes, but the main species giving rise to taint are volatile materials exemplified by those compounds shown in Table 2.1. However, bitter compounds also give rise to the perception of taints and off-flavours. These are frequently water-soluble involatile materials that are detected by the sense of taste.

The complex nature of food quality perception creates many difficulties for the sensory analyst, whose primary task is to use human subjects as an instrument to measure the sensory quality of foods. The perceptual processes described above are determined by physiological responses to stimuli, but when using human subjects for sensory analysis, psychological aspects must also be considered. The factors that should be considered in assessing the performance of human subjects in this way are accuracy, precision and validity (Piggott, 1995). Sensory measurements are a direct measure of human response. They have an inherently higher validity than instrumental measures, which are nonetheless of value as a complement to sensory data, for example in shelf-life assessment. In measuring human responses, low precision must be expected, but variation can be reduced by careful selection of a range of human subjects who can produce a response with lower variability, and by extensive training.

Improving accuracy (giving the correct answer without systematic error or bias) can be achieved by recognising the various sources of physiological and psychological biases that can influence human subjects. The effect of physiological differences between individuals can be reduced, but not completely eliminated, by careful selection procedures. Psychological factors can introduce systematic biases that might not be recognised. These include those arising from unwanted interaction between panellists, and those from more subtle sources. These can be greatly reduced by choice of sensory test procedure and by careful experimental design and operation of sensory test procedure.

2.1.2 Thresholds

A common means of quantifying response to chemical stimuli is through the use of a threshold, commonly defined as the concentration in a specified medium that is detected by 50% of a specified population. This definition is widely used in describing sensory perception of stimuli, but unfortunately is frequently misused and misunderstood. Thresholds indicate the level of stimulus that is sufficient to trigger perception but, contrary to common usage, a number of thresholds can be defined, none of which is invariant. Standards give the following definitions for thresholds relevant to taint testing (BSI, 1992):

- **Detection threshold:** the lowest physical intensity at which a stimulus is perceptible
- **Recognition threshold:** the lowest physical intensity at which a stimulus is correctly identified.

In dealing with taints, we are generally concerned with detection thresholds, but usually literature data do not define whether detection or recognition thresholds are being quoted. Literature data also frequently fail to cite the methodological variables, such as number of test subjects, degree of experience of test subjects, nature of instructions to test subjects, test procedure and whether replicated, and details of any statistical analysis. These omissions may serve to explain the wide range of numerical values found by different researchers for the same thresholds.

Even if threshold measurements utilised the same test methodologies and exercised careful control over experimental variables, variations in measured thresholds must be expected as a result of the enormous range of human sensitivities. Measurements should therefore use as many human subjects as possible and should also include provision for repeat testing, since subject performance improves with practice. The medium in which the stimulus is present has a substantial effect on the measured thresholds, through masking effects from other flavours and from the different rate and extent of release that can occur.

2.2 Sensory test procedures

Any high-quality sensory evaluation system needs to satisfy a number of inter-related requirements. These are discussed below and more detailed discussions can be found in standard texts (e.g. Piggott, 1988; Meilgaard *et al.*, 1999; Stone and Sidel, 1993; Lawless and Heymann, 1998). The requirements are shown schematically in Fig. 2.2.

2.2.1 Objectives

Clear objectives are central to the establishment of any system that will be sufficiently accurate to measure the required sensory characteristics with

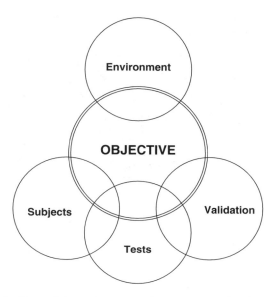

Fig. 2.2 Essential requirements of a sensory evaluation system.

the required precision and that will be cost-effective. Testing for taint is most appropriately carried out as part of a quality control function, although some companies (for example, suppliers of packaging to the food industry) may need to carry out research investigations. The objectives of the system will then be defined by the quality procedures operated by the company.

2.2.2 Test environment

A suitable environment is essential for generating high-quality sensory data with minimal bias. The environment is important not only in providing standardised working conditions for the assessors, but also in providing a work area for sample preparation and for data analysis. Detailed advice is given in a number of publications (e.g. Stone and Sidel, 1993; BSI, 1989). The three main components of a sensory evaluation environment are:

- a preparation area of adequate size and appropriately equipped
- a testing environment, adjacent to, but separated from, the preparation area
- individual booths to eliminate assessor interaction.

2.2.3 Type of test and test subjects

Many sensory test methodologies are available, but fall into two main classes, shown schematically in Fig. 2.3:

12   Taints and off-flavours in food

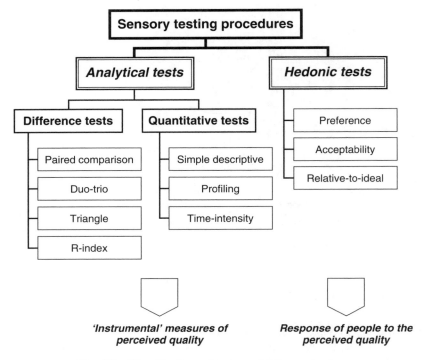

Fig. 2.3 Classification of sensory testing procedures.

- *Analytical tests*. These tests are used to measure sensory characteristics of products by providing answers to the questions:
  - is there a difference?
  - what is the nature of the difference(s)?
  - how big is (are) the difference(s)?
- *Hedonic/affective tests*. These tests are used to measure consumer response to sensory characteristics of the products by providing answers to the questions:
  - which product is preferred?
  - how much is it liked?

The two classes comprise tests that satisfy completely different objectives and which are subject to different operating principles. Analytical tests use human subjects as a form of instrument to measure properties of the food. Hedonic tests measure the response of consumer populations to the food in terms of likes or dislikes. Different psychological processes are used for each type of test. In general there is no simple linear relationship between the two types of data. It is of great practical importance that the type and numbers of subjects used for the analytical and hedonic tests are quite different.

The subjects to be used are defined by the objective of the test and by the consequential choice of test. The numbers of subjects to be used depend on the level of expertise and training of the assessors. Recommended minimum numbers are give in BS 5929 Part 1, 1986 (ISO 6658) (BSI, 1986), which also discriminates between assessors, selected assessors and experts.

2.2.4 Analytical tests

Both discriminative and descriptive tests use small panels of assessors chosen for their abilities to carry out the tests. Guidelines for establishing such assessors are given in BS 7667 Part 1, 1993 (ISO 8586-1) (BSI, 1993). A general scheme for establishing a panel requires the following steps:

- *Recruitment.* Panellists can be recruited from within the company, or dedicated part-time panellists can be recruited from the local population (company employees should not be compelled to participate).
- *Screening.* These preliminary tests are used to establish that sensory impairment is absent, to establish sensitivity to appropriate stimuli and to evaluate the ability to verbalise and communicate responses. These tests will depend mainly on the defined objectives of the sensory testing, but will typically consist of the following:
    - the ability to detect and describe the four basic tastes: sweet, sour, salt and bitter; these may be extended to cover metallic, umami and astringent.
    - the ability to detect and recognise common odorants, together with those characteristic of the product range of interest.
    - the ability to order correctly increasing intensities of a specific stimulus, for example increasing sweetness or increasing firmness.
    - the ability to describe textural terms characteristic of relevant food types.
    - absence of colour vision deficiencies, (approximately 8% men, but only 0.4% women suffer colour vision deficiencies), tests can be carried out using Ishihara charts (available from opticians or booksellers).

    Selection of suitable panellists is usually made on the basis of a good performance across the entire range of tests, rather than excellence in some and poor response to others. If the panel is to be used for a specific purpose, then the tests relevant to that purpose can be weighted appropriately.
- *Training.* In the initial stages, training is limited to the basic principles and operations, following which further selection can be made. More closely targeted training can then be carried out using the products of interest and aimed towards the specific tests to be used in practice.

- *Monitoring.* Close monitoring of panel performance is essential and any drift that is identified must be corrected by retraining procedures.

2.2.5 Hedonic tests

Subjects (respondents) for hedonic tests are chosen to represent the target consumer population and to reflect any inhomogeneity in that population. Consequently, they need to be used in sufficient numbers to give statistical confidence that they are representative. They must be given the opportunity to behave as they would in a real consumption environment. In particular, they must not be selected on the basis of sensory ability and must not be given any training. Numbers in excess of 100 respondents are normally used. For the early stages of concept development, qualitative studies using focus groups with small numbers of respondents can be used, but the data generated should be treated carefully and conclusions must not be generalised. The same subjects must not be used for both types of test and, in particular, in-house staff must not be used to generate hedonic data that may be viewed as consumer related.

2.3 Types of test: difference/discrimination tests

Difference or discrimination tests are perceived as one of the easiest classes of sensory testing to apply in an industrial environment and are consequently heavily used. The tests can be used in two ways, either to determine whether there is an overall difference between two samples or to determine whether one sample has more or less of a specific attribute than another. The tests have limited information content and can be unwieldy when many product comparisons are to be made. In such circumstances, alternative methods, such as profiling, are often superior, but the high sensitivity of well-designed difference tests can offer the best protection against taint problems. Difference tests are almost universally used to ascertain whether two samples are different, not to ascertain whether two samples are the same. It should be noted that, if a difference is not found, it does not prove that samples are the same. However, future revisions of ISO standards will advise sensory analysts on how to use the tests for the latter purpose.

2.3.1 Paired comparison test

In the most common form of the test, two coded samples are presented either sequentially or simultaneously in a balanced presentation order (i.e. AB and BA). There are two variations in the test. In the directional difference variant, the panellists are asked to choose the sample with the greater or lesser amount of a specified characteristic. The panellists are usually instructed to make a choice (forced choice procedure), even if they have to make a guess, or they may be allowed to record a 'no-difference' response. In the directional form test (less commonly referred to as the 2-AFC (alternative forced choice) test), it is important that the panellists clearly comprehend the nature of the attribute of interest. It can be argued that if time is needed to train panellists to recognise a specific characteristic, a descriptive test should have been selected.

2.3.2 Duo–trio test
In the most common variant of the duo–trio test, the panellists are presented with a sample that is identified as a reference, followed by two coded samples, one of which is the same as the reference and the other different. These coded samples are presented in a balanced presentation order, i.e.

A (reference)   A   B
A (reference)   B   A

The panellists are asked to identify which sample is the same as the reference. The duo–trio test is particularly useful when testing foods that are difficult to prepare in identical portions. Testing such heterogeneous foods using the triangle test, which relies on identical portions, can give rise to difficulties, but in the duo–trio test there are no inherent difficulties in asking the question 'Which sample is most similar to the reference?'.

2.3.3 Triangle test
Three coded samples are presented to the panellists, two of which are identical, using all possible sample permutations, i.e.

ABB   AAB
BAB   ABA
BBA   BAA

The panellists are asked to select the odd sample in either fixed-choice (preferred) or no-difference procedures. The increased number of samples compared with a paired comparison test can result in problems with flavour carry-over when using strongly flavoured samples, making identification of the odd sample more difficult. Difficulties can also be encountered in ensuring presentation of identical samples of some foods.

2.3.4 3-AFC (alternative forced choice) test
A less common procedure uses one-half of the same sample permutations from the triangle test in a triad format, but either the difference of interest between the samples is revealed to the panellists in advance, or the

panellists identify the nature of any difference in advance. In the test itself, the panellists are then asked to identify the sample (or samples) with the specified characteristic. For example, a typical instruction might be 'One of these samples is more bitter than the others; please identify this sample'. O'Mahony (1995) has identified the reasons why this test can be more sensitive than the triangle test, but the test suffers from the need to identify the nature of the difference positively in advance.

2.3.5 R-index test

The short-cut signal-detection method of the R-index test is a relatively recent development (O'Mahony, 1979; 1986) and is less well used, but an application to taint testing has been described (Linssen et al., 1991). The test samples are compared against a previously presented standard and rated in one of four categories. For difference testing, these categories are standard, perhaps standard, perhaps not standard and not standard. The test can also be carried out as a recognition test, in which case the categories are standard recognised, perhaps standard recognised, perhaps standard not recognised and standard not recognised. The results are expressed in terms of R-indices, which represent probability values of correct discrimination or correct identification. The method is claimed to give some quantification of magnitude of difference, but its use has not been widely reported in the literature. One important limitation is that a relatively high number of judgements is needed in this form of test, leading to the risk of severe panellist fatigue, and in the case of some important taints, severe sensory adaptation that can result in non-identification.

2.3.6 Difference from control test

The test is of particular value when a control is available; the panellists are presented with an identified control and a range of test samples. They are asked to rate the samples on suitable scales anchored by the points 'not different from control' to 'very different from control'. The test results are usually analysed as scaled data.

2.4 Types of test: quantitative and hedonic tests

The major advantages of discrimination tests are their relative simplicity to set up and operate, and their high sensitivity. However, they have two important limitations. First, only two sample treatments are compared together. Second, the information content of discrimination tests is limited, even when operated in an extended format, incorporating a range of questions. More informative tests can produce more quantitative data, which can be subjected to a wider range of statistical treatments.

2.4.1 Scaling procedures

Quantification of sensory data is needed in many applications. The recording of perceived intensity of attributes or liking requires some form of scaling procedure. These procedures should be distinguished from quality grading systems, which are used to sort products into classes defined by a combination of sensory characteristics. Such systems are not open to quantitative numerical analysis. Scaling procedures are mainly used to generate numeric data that can be manipulated and analysed statistically. Before this can be carried out, however, thought must be given to how the scales used are seen and interpreted by the assessors and how this may influence the type of analysis that can be safely applied. The different types of scale used are described below:

- *Category scales* use a defined number of boxes or categories (often 5, 7 or 9, although other numbers are sometimes used). The scale ends are defined by verbal anchors and intermediate scale points can be given verbal descriptions.
- *Graphic scales (line scales)* consist of a horizontal or vertical line with a minimum number of verbal anchors, usually at the ends. Other anchors can be used, for example, to define a central point, or to denote the position of a reference sample.
- *Unipolar scales* have a zero at one end and are most commonly used in profiling, especially for flavour attributes.
- *Bipolar scales* have opposite attributes at either end. Definition of the central point can often give rise to logical difficulties, as can ensuring that the extreme anchors are true opposites. This can be a particular problem for textural attributes, for example, when using 'soft' to 'hard' type scales. Bipolar scales are frequently used for consumer acceptability testing, especially using the 'like extremely' to 'dislike extremely' format.
- *Hedonic scales* are used to measure consumer liking or acceptability. Category scales are usually used.
- *Relative to ideal scales* are a type of hedonic scale, which measures deviation from a personal ideal point.

The type of scale used and its construction depend on a number of factors:

- *Purpose of test*. Both category and graphic scales are commonly used with trained panels. In consumer testing, category scaling methods are usually used.
- *Expertise of assessors*. Untrained assessors are generally poor discriminators and can discriminate over only a small number of scale points. Consequently, 5-point or even 3-point category scales are often used with consumers. Trained panels can start with 5-point or 7-point category scales, but as their discrimination ability increases, they can use effectively more scale points or graphic scales. When using

inexperienced assessors or consumers, scales incorporating a 'neutral point', such as the central point in an odd-numbered category scale, are sometimes avoided in order to minimise the risk of 'fence-sitting'.
- *Number of assessors.* Using small assessor numbers with a low number of category scale points will limit statistical analysis options.
- *Data-handling facilities.* Category scaling responses can be entered relatively quickly onto a spreadsheet, whereas data from line scales must be measured, which can be a time-consuming procedure. Computerised data acquisition, either directly from a terminal or indirectly from optical readers, can avoid this problem.

In practice, establishing a trained sensory panel can often proceed from a category scale with a small number of scale points (e.g. five), through a category scale with more points (e.g. nine) to a line scale. Sensory analysts should be aware of difficulties that panellists have in using scales and careful training is needed to ensure that scales are unambiguous and can measure the intended response.

2.4.2 Simple descriptive procedures

Scaling may often be needed in order to quantify a single well-defined attribute. However, it should be established that there is no ambiguity in the attribute of interest. This is particularly relevant during product development or modification, when the assumption that a process or ingredient modification will change only a single attribute is frequently violated. Such changes are especially common when textural changes are a consequence of process or ingredient modifications. If it is suspected that several attributes might be of interest, then the profiling procedures described in the subsequent sections should be considered.

2.4.3 Sensory profile tests

Profile tests are a means of quantifying the set of sensory attributes of foods that have been established using descriptive tests. Several types of profiling methods are available, but the most commonly used procedures are based on quantitative descriptive analysis (QDA), which uses a list of attributes generated by the panellists for the food type concerned. Following agreement of attributes and their definitions, suitable scoring scales (usually unstructured line scales) are constructed with appropriate anchors. The panellists are trained in scoring reproducibly the intensity of those attributes using a training set of samples. Training can be lengthy for foods characterised by large attribute numbers. Once satisfactory reproducibility has been achieved, the test samples are scored in replicated tests using appropriate statistical designs. The data are analysed statistically using appropriate methods, for example, analysis of variance and multiple difference testing, but multivariate analysis methods such as principal component analysis are increasingly common. Graphical methods are often used to display results.

An alternative profile method is free-choice profiling, in which the assessors use the individual attribute list generated as the first stage of a simple qualitative test. Each assessor constructs an individual profile based on his/her own attributes and a consensus profile is constructed mathematically using a technique known as generalised Procrustes analysis. This procedure reduces panel training times considerably, but the consensus data are difficult to interpret reliably.

2.4.4 Time-related methods

Sensory attributes are not perceived instantaneously and can change in intensity with time in the mouth. In particular, the intensity of off-characteristics such as bitterness can develop with time. Time–intensity methods are used to measure intensity of a specific attribute as a function of time in the mouth. They have been used extensively to investigate the temporal behaviour of tastants, such as sweet and bitter molecules, and the release of volatile flavour materials from foods (Overbosch et al., 1991; Shamil et al., 1992). The use of time–intensity for flavour measurement is relatively well established and textural changes can also be monitored using the method.

A major limitation of the time–intensity method is that only a single attribute can be tracked with time and if a number of important attributes are thought to be time-dependent, separate sessions are needed for each attribute. Difficulties encountered in time–intensity profiling prompted the development of a hybrid technique, progressive profiling (Jack et al., 1994). In this technique, assessors carried out a profile on a set of texture descriptors at each chew stroke over the mastication period. Such a method has a number of potential advantages: several attributes can be assessed in one session, scaling is reduced to a unidimensional process and the most important aspects of the shape of a time–intensity curve are retained.

2.4.5 Consumer acceptability testing

Consumer tests give a direct measure of liking that can be used more directly to estimate shelf-life. The most common procedure is to ask consumers representative of the target population to scale acceptability on a 9-point category scale, anchored from 'like extremely' to 'dislike extremely'. A minimum of 50 consumers should be used and preferably 100 or more for increased statistical confidence. Suitable experimental designs should be used and appropriate statistical analysis. Other information on individual modalities (appearance, odour, flavour and texture) can also be obtained, together with attribute intensity information, but it is preferable to keep such tests simple and to focus on overall acceptability. The most common procedure for operating the tests is to recruit consumers from a convenient high street or mall location and to carry out the tests in a convenient hall. Alternatively, a mobile test laboratory can be used to increase the degree of control.

2.5 Data handling, analysis and presentation

Sensory experiments can generate large amounts of data and reliable conclusions require validation using statistical techniques. Details of suitable statistical methods can be found in a number of texts, e.g. O'Mahony (1986), Smith (1988), Meilgaard *et al.* (1999) and Lahiff and Leland (1994). Different types of sensory test procedure generally utilise specific analysis procedures. Analysis of difference tests usually requires comparison of the test result against tables generated using the binomial expansion (e.g. Stone and Sidel, 1993). In the case of the more sophisticated profiling techniques, a wide range of options is available, both univariate and multivariate. Many statistical software packages are now available. The most sophisticated require a sound understanding of statistical principles, but more user-friendly packages are available that satisfy most requirements. However, it is usually found that no single package can cover the entire range of basic requirements. Clear and effective presentation of sensory data, including the results of statistical tests, is essential. Most standard spreadsheets are now able to offer a wide range of presentation possibilities for both univariate and multivariate data. Specific aspects of the analysis of taint data are covered in Section 2.6.2.

2.6 Choosing and interpreting sensory tests

Sensory testing is employed extensively in the food and drinks industries for many purposes. Taint testing is commonly carried out as part of routine quality control procedures, but other applications can be seen for other purposes, for example in the selection of packaging materials in product development programmes.

2.6.1 Test selection and operation

Selecting sensory testing procedures for taint testing encounters a fundamental problem. A taint that could spell commercial disaster may be detectable only by a few per cent of consumers. So can sensory tests that, for practical reasons, use only small numbers of panellists, be designed to guard against this occurrence? Although no procedure of practical value can guarantee that a taint will be detected, steps can be taken to minimise the risk of not identifying a taint stimulus. The most important of these are the following:

- For all test procedures, if the identity of the tainting species to be tested for is known, use panellists who are known to be sensitive to that species. Unfortunately, it cannot be assumed that a panellist sensitive to one specific tainting species will also be sensitive to other tainting species.
- If a high-sensitivity panel is not attainable, especially if the nature of the taint is unknown, use as many panellists as possible in the hope of having someone present who is sensitive to the taint. Practical constraints will limit the number used, but if possible, this should not be less than 15. There is little value in using a smaller number repeatedly in replicated tests if their sensitivities are low.
- Use a high-sensitivity test procedure. Difference tests are generally more suitable than profile-type tests, as they are more rapid and do not require intensive training. In addition, a difference test against an appropriate, untainted control is a relatively easy task for the panellist. Triangular tests are commonly used, but if there is a risk of flavour carry-over, a duo–trio test using an untainted reference should be considered. Other sensory tests have shown few applications in taint testing, but for rapid screening of a relatively large number of samples, scaling of taint intensity on either a category scale or an unstructured line scale can be of value.
- When using difference tests, maximise the information content of the test by using an extended format. A rigorous approach to sensory analysis would dictate that identification of a difference is the only information that should be elicited from panellists, the reason being that any attempt to elicit other information will require different mental processes that may invalidate the test. As a minimum requirement, descriptive information on the nature of any identified difference must be recorded. In addition, at Leatherhead Food International, two other types of information are elicited. First, since taints are by definition disliked, preference information is recorded. As indicated previously, this is a unique exception to the general rule that hedonic and analytical tests must not be mixed. The preference information is not interpreted as a likely measure of consumer response, but is used purely as a directional indicator in conjunction with descriptive information. Second, panellists are asked to rate how confident they were in their choice of the odd sample on a 4-point category scale (absolutely sure/fairly sure/not very sure/only guessed). Confidence levels weighted toward one end of the scale or the other can help resolve indeterminate results by indicating to what extent panellists may be guessing. Such a scale may

be formalised by assigning scores to the scale points. An important point to note when using such ancillary data, however, is that these data are valid only from panellists who have correctly identified the odd sample. Data from panellists who have made incorrect identifications are invalid and must not be used.

2.6.2 Analysis of test data

It is essential when analysing the results of sensory taint test data to minimise the risk of not identifying a taint that is present and to use statistical tests that are appropriate in this context. A fundamental problem is apparent here, as discussed by O'Mahony (1982, 1986). Conventional hypothesis testing involves testing the experimental data against a null hypothesis (H0) that no trend, or difference, exists in the data. A probability value is calculated that represents a difference occurring by chance. If this value is low, it is unlikely that the null hypothesis is true, and the alternative hypothesis (H1) is accepted, which states that a difference is present. On the other hand, a high value indicates that the result could have occurred by chance and the null hypothesis is not rejected. A probability value of 0.05 (5% significance) in a difference test can then be interpreted as indicating that a difference does appear to exist, but with a 5% (1 in 20) probability that the result could have been due to chance. If we require more assurance that we really have found a difference, a lower significance level of 1% could be used, giving a 1 in 100 probability of a chance result.

Unfortunately, the more assurance of a real difference that we seek, the greater the risk of not identifying a real difference that is present (Type II error). By increasing the significance level to 10%, 15% or even 20%, the risk of not identifying a real difference diminishes, but the risk of incorrectly identifying a difference (Type I error) increases. The choice of an appropriate cut-off point depends on how prepared you are to be wrong; even 1% would be too high a risk in medical experiments, where values of 0.1% or 0.01% may be more appropriate. In sensory testing, however, and in particular in taint testing, the consequences of incorrectly saying that a difference exists are relatively minor, against the consequences of not identifying a difference and allowing tainted product to reach consumers. Consequently, levels of up to 20% should be used to minimise this risk, but accepting that by using a 20% cut-off, there will be an expectation that overall 1 in 5 will be incorrect.

It should be noted that in interpreting probability levels, there is little practical difference between probabilities of 4.9% and 5.1%, but that, if a rigid cut-off of 5% were used, different interpretations would result. Consequently, it is preferable to calculate exact significance values and use common sense in their interpretation. Regardless of the results of statistical tests, take careful note of minority judgements, particularly from panellists of established reliability, and retest for added assurance.

2.7 Applications of sensory testing

Sensory testing procedures are employed in the context of taints and off-flavours in two different ways. First, diagnostic sensory testing can be carried out to investigate or confirm the presence of taint and to generate descriptive information that can be used to assist chemical analysis. Second, sensory testing can be carried out as part of preventive procedures designed to determine the risks carried by food-contact materials and other environmental factors in transferring tainting materials to food.

2.7.1 Diagnostic taint testing

Taint problems continue to be widespread in spite of considerable effort and expense on the part of the food and associated industries. These problems frequently involve insurance claims or litigation, and in such cases correct sensory (and also chemical analysis) procedures must be adhered to rigorously.

The first indication of a taint problem is usually through consumer complaints on sensory quality. One consequence of the commonly low level of taint detection is that the complaints may come in at a low rate over a period of time and recognition of a taint problem may not be immediate. In addition, investigation of a sensory quality complaint arising from a single customer return requires care, owing to possible safety problems. Examination should be restricted to odour and, if feasible, chemical composition.

Examination of batches of suspect product should be carried out as a means of investigation but, again, care must be taken to guard against possible safety problems. The suspect product to be tested should be from the same batch coding as the complaint material and as far as possible should have gone through the same distribution channels. In addition, suitable control material of similar age should be available. Availability of retained samples from points in the production and distribution chain is invaluable. In circumstances in which the complaint pattern suggests non-uniform distribution within a production batch, testing can be carried out to a suitable statistical sampling plan, but such testing can often prove prohibitively time consuming and expensive.

Consumer descriptions of most taints cannot be relied on as a means of focusing chemical analysis investigations, so sensory testing of suspect batches should be carried out to generate reliable descriptive information. However, care is needed in relating descriptions to possible chemical species. If the presence of a taint in complaint batches can be established, efforts must be made as quickly as possible to isolate affected product and to identify the source of the taint. Sensory testing can be used to investigate whether the problem is associated with a single transport container, production run, ingredients batch or packaging material batch. If the problem appears to be continuing over a period of time, however, possible sources such as new building materials, process line components or waterborne contamination must be examined. If ingredients (including water supply) are suspected as continuing sources of taint, small test batches of product can be prepared and compared against appropriate controls. Materials suspected as sources of taint can be tested using taint transfer tests, as described in the next section.

Particular care must be taken in gathering evidence and setting up test procedures if, as must frequently be assumed, insurance claims are likely or, even more importantly, litigation is likely. Companies supplying tainted materials may face litigation by their customers, and in turn may enter into litigation against their own suppliers. It is frequently advantageous to contract out testing work to an experienced third-party organisation in order to establish impartiality in generating data to be used as evidence. Care should, however, be taken to establish the scientific credentials and expertise of such organisations. A number of suggestions can be made in initiating such investigations if the timescales and costs of litigation are to be minimised:

- Have in place documented systems for rapid identification of the nature and source of the taint.
- Isolate affected product batch codes.
- Use both sensory and chemical analysis to establish both the occurrence and the identity of the taint – do not rely on one type of information only.
- Store both suspect and control samples under conditions suitable for future testing.
- Carry out sensory testing according to international standards procedures and use as many assessors (preferably sensitive) as possible.
- Extract as much information from the tests as possible, but do not compromise the test quality.
- Have the tests carried out and interpreted on a double-blind basis, especially if the tests are to be subcontracted to a third-party organisation.
- Ensure that the names and addresses of panellists are held, as presentation of sensory data in a court of law may require the presence of the individual panellists as witnesses.

2.7.2 Preventive testing (taint transfer)

Preventive testing is a powerful, but frequently misapplied, means of limiting problems arising from the introduction of new materials and changes in environmental conditions. The tests seek to expose food or food simulants to potential taint sources in an exposure situation that is severe but not unrealistic. Severity factors of up to ten times are usually used, but

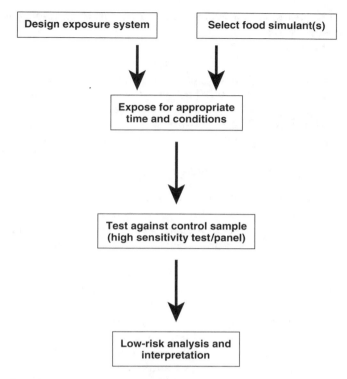

Fig. 2.4 Protocol for setting up and operating taint transfer tests.

higher factors can be used for critical applications. An outline protocol for such tests is shown in Fig. 2.4.

The design of the exposure system varies considerably depending on the nature of the test. For example, taint testing of pesticide residues requires a full-scale field trial with rigidly defined crop growing, pesticide application and crop sampling procedures. In testing packaging systems, the model system may need to simulate either direct contact or remote exposure and, in testing process line components, factors such as product residence time and product temperature must be considered.

General factors to be considered in designing model systems for testing materials such as flooring, paints and packaging materials include the following:

- type of food/food simulant
- ratio of the volume or surface area of the material to the volume of the vessel
- ratio of the volume or surface area of the material to the volume or surface area of the food/food simulant

26 Taints and off-flavours in food

- stage of exposure (e.g. at what stage during curing of a flooring material exposure is to start)
- length of exposure
- temperature and humidity at exposure
- exposure method (e.g. direct contact or vapour phase transfer)
- exposure lighting conditions (especially when rancidity development may occur)
- ventilated or unventilated exposure system
- temperature and length of storage of food/food simulant between exposure and testing
- sensory test procedure and interpretation.

Choice of appropriate foods/food simulants is an important consideration, with two possible approaches. Where a specific ingredient or product is known to be at risk, the test can be specifically designed around it. Where the purpose of the test is more general, however, simple foods or food simulants are often used. Solvent or adsorptive properties are the most important physicochemical considerations in selecting appropriate general simulants. Oils and fats will tend to absorb water-insoluble tainting species, and materials such as butter are known to be sensitive to taint transfer. High surface area powders with hydrophilic characteristics have also been found to be sensitive to taint transfer and tend to absorb water-soluble taints. Use of such materials will simulate a large proportion of the solvent and adsorptive characteristics of real foods. An additional requirement for suitable simulants, however, is that they should be relatively bland to enable easy detection, and also of acceptable palatability. This latter consideration, unfortunately, renders some simulants recommended for packaging migration tests, for example 3% acetic acid, unsuitable for taint transfer testing. Still mineral water can be used to simulate aqueous liquids, and 8% ethanol in water to simulate alcoholic drinks. In the author's laboratory, however, the characteristic ethanol flavour has been found to be rather unpleasant, and a bland vodka has been used, diluted down to 8% ethanol. Some suitable materials for general purpose use are given in Table 2.3.

2.8 Standardisation of test methods and instrumental methods

Standard procedures for taint transfer testing have been published in several countries, mainly aimed at food packaging materials (for example, BSI, 1964; OICC, 1998; DIN, 1983; ASTM, 1988). The British Standard and the American Standard deal with taint transfer from packaging films in general, and the OICC Standard ('Robinson test') deals specifically with taint transfer to cocoa and chocolate products, although it is frequently used for other products. The German DIN Standard also refers to food packag-

Table 2.3 Foods/food simulants for taint transfer testing

Type	Food/simulant	Comments
Fat	Chocolate	Bland variety (white or milk)
	Unsalted butter	Mixed prior to sensory testing, or outer surfaces only used for severe test
Hydrophilic powder	Sugar	High surface area preferred (e.g. icing sugar); test as 5% solution
	Cornflour	Test as blancmange formulation (can get textural variation)
	Rusks/crispbread	Expose crushed
Combined	Biscuits	High fat, e.g. shortbread
	Milk	Full cream; for short-term exposure tests only, or rancidity problems can interfere

ing, but contains much useful information for setting up tests on other materials. All the published methods are, however, deficient in their use of sensory testing methods, and the more correct and more sensitive procedures described here and based on BS 5929 (BSI, 1986) should be used to replace these where possible. At the time of writing, more comprehensive standards for testing packaging materials are under development.

In seeking to maintain high food quality and to minimise the risk of taint problems, the UK retailer Marks & Spencer has developed codes of practice referring to the use of packaging films, plastics and paints (Goldenberg and Matheson, 1975). These guidelines stress the importance of testing by the packaging supplier before dispatch and by the food manufacturer before use. This important principle is, unfortunately, rarely recognised by the food industry in general. Food manufacturers frequently rely on suppliers to provide some general form of certification or test evidence that a material is free from taint, but the material is seldom tested under the conditions in which it will be used. Information provided by suppliers can be regarded as useful screening information, but users must protect themselves by re-testing under more realistic and rigorous conditions.

Sensory testing is an essential component of taint identification, but alone seldom gives a positive identification of a chemical species. Instrumental methods of chemical analysis must be used for positive identification, especially in situations in which litigation or insurance claims occur. As most tainting species are volatile, gas chromatography–mass spectrometry is commonly used, but with prior extraction/concentration steps in order to increase sensitivity. Identification of involatile materials, for example those responsible for unwanted bitter flavours, can be a more difficult problem, especially as sensory descriptions are imprecise and rarely helpful.

2.9 Ethical aspects

Any sensory evaluation operation using human subjects as a means of acquiring information on the sensory characteristics of foods must have ethical procedures in place designed to protect panellists from hazards associated with consuming unsafe food, and these must form part of general safety practices operated by the company management. Consuming or testing food that may be contaminated with unknown tainting species carries a specific toxic risk, and additional measures may be needed to protect panellists against such risks and also company staff against subsequent litigation. Companies are now increasingly using guidelines originally drawn up in the UK for testing genetically modified ingredients (ACNFP, undated), but which are now regarded as being applicable to any situation in which there is a potential toxicity hazard associated with the ingestion of chemical substances, including novel ingredients. The guidelines define an exposure/risk matrix; depending on the information known about the test material, the position within the matrix defines whether or not human testing should take place. Rigid application of these guidelines would create difficulties in taint testing owing to the unknown nature of taints, but they also stress the overriding need for any organisation carrying out testing using human subjects to have a documented ethical procedure designed to protect human subjects, if necessary utilising external expertise. An essential component of any ethical system is an advise and consent procedure, in which all known risks are summarised and panellists are allowed to withdraw from the test. Information that should be given includes:

- type of risk
- chemical name of any contaminants, if known
- natural occurrence in foods, and normal use (if appropriate)
- available toxicity data, either in numeric terms (e.g. LD_{50}) or relative to appropriate materials more familiar to the panellists
- approximate quantity that may be ingested during a test and anticipated number of tests over the course of the projects; if appropriate, panellists may be instructed to expectorate the samples and not to swallow, but some involuntary swallowing must be expected.

2.10 Future trends

Assuring the sensory quality of foods is a goal for the entire food industry, but until recently detailed specifications for food quality have relied almost entirely on non-sensory factors. A potential adaptation of sensory methods, driven by the retail sector in the UK, is the development of detailed sensory specifications for foods and incorporates a simple assessment of product quality against specification. Although relatively crude, such systems offer the opportunity for low-cost sensory appraisal of perceived quality on a qualitative or semi-quantitative basis and should be of great assistance in ensuring freedom from taint.

The development in instrumental methods is likely to follow the route exemplified by the 'electronic nose' systems, more correctly described as volatile sensors (Schaller *et al.*, 1998). At present, these systems are detection instruments, which cannot easily identify specific volatiles, although more recent instruments are more correctly regarded as developments of mass spectrometers. However, they are more usefully used as pattern recognition devices, using multivariate or neural network software systems. These can detect changes in volatile patterns that can potentially be related to foreign volatile components. Recent publications have shown potential applications in the detection of taint in cork stoppers (Rocha *et al.*, 1998) and of boar taint in pork (Annor-Frempong *et al.*, 1998). Unconfirmed reports have indicated that some food manufacturers are now using electronic noses to screen incoming packaging materials for odour level. Recent reports have also indicated that similar sensing and pattern recognition systems could also be used for involatiles, although these may be less relevant to storage changes (Lavigne *et al.*, 1998).

2.11 References

ACNFP (undated), 'Guidelines on the conduct of the taste trials involving novel foods or foods produced by novel processes', Advisory Committee on Novel Foods and Processes, London.

ANNOR-FREMPONG I E, NUTE G R, WOOD J D, WHITTINGTON F W and WEST A (1998), 'The measurement of the responses to odour intensities of "boar taint" using a sensory panel and an electronic nose' *Meat Sci*, **50**(2), 139–51.

ASTM (1988), ASTM E619-84 'Evaluating foreign odors in food packaging', American Society for Testing and Materials, Philadelphia.

BRITISH STANDARDS INSTITUTION (1964), British Standard BS 3755 'Methods of test for the assessment of odour from packaging materials used for foodstuffs', London.

BRITISH STANDARDS INSTITUTION (1986), British Standard BS 5929 Part 1; ISO 6658. 'Introduction and general guide to methodology', London.

BRITISH STANDARDS INSTITUTION (1989), British Standard BS 7183; ISO 8589. 'Guide to design of test rooms for sensory analysis of food', London.

BRITISH STANDARDS INSTITUTION (1992), British Standard BS 5098; ISO 5492. 'Glossary of terms relating to sensory analysis', London.

BRITISH STANDARDS INSTITUTION (1993), British Standard BS 7667 Part 1; ISO 8586-1. 'Assessors for sensory analysis. Part 1. Guide to the selection, training and monitoring of selected assessors', London.

DEUTSCHES INSTITUT FÜR NORMUNG (1983), DIN 10955 'Testing of container materials and containers for food products', Berlin.

GOLDENBERG N and MATHESON H R (1975), '"Off-flavours" in foods, a summary of experience: 1948–74', *Chem Ind*, 551–7.

HUTCHINGS J B (1994), *Food Colour and Appearance*, Blackie A&P, Glasgow.

JACK F R, PIGGOTT J R and PATERSON A (1994), 'Analysis of textural changes in hard cheese during mastication by progressive profiling', *J Food Sci*, **59**(3), 539–43.

LAHIFF M and LELAND J V (1994), 'Statistical methods', in *Source Book of Flavors*, G. Reineccius (ed), Chapman & Hall, New York, 743–87.

LAVIGNE J J, SAVOY S, CLEVENGER M B, RITCHIE J E, YOO S-Y, ANSYLN E V, MCDEVITT J T, SHEAR J B and NEIKIRK D (1998), 'Solution-based analysis of multiple analytes by a sensor array: Toward the development of an "Electronic tongue"', *J Am Chem Soc*, **120**, 6429–30.

LAWLESS H T and HEYMANN H (1998), *Sensory Evaluation of Food. Principles and Practices*, Chapman & Hall, London.

LINSSEN J P H, JANSSENS J L G M, REITSMA J C E and ROOZEN J P (1991), 'Sensory analysis of polystyrene packaging material taint in cocoa powder for drinks and chocolate flakes', *Food Additives Contamin*, **8**(1), 1–7.

MEILGAARD M, CIVILLE G V and CARR B T (1999), *Sensory Evaluation Techniques*, 3rd edn, CRC Press, Boca Raton, Fl.

OFFICE INTERNATIONAL DU CACAO ET DU CHOCOLAT (OICC) (1998), 'Transfer of packaging odours to cocoa and chocolate products', Analytical Methods of the Office International du Cacao et du Chocolat, Verlag, Zurich.

O'MAHONY M A P D (1979), 'Short-cut signal detection measures for sensory analysis', *J Food Sci*, **44**, 302–3.

O'MAHONY M A P D (1982), 'Some assumptions and difficulties with common statistics for sensory analysis', *Food Technol*, **36**(11), 76–82.

O'MAHONY M A P D (1986), *Sensory Evaluation of Food: Statistical Methods and Procedures*. Marcel Dekker, New York.

O'MAHONY M A P D (1995), 'Who told you the triangle test was simple?' *Food Qual Preference*, **6**(4), 227–38.

OVERBOSCH P, AFTEROF W G M and HARING P G M (1991), 'Flavour release in the mouth', *Food Rev Internat*, **7**, 137–84.

PIGGOTT J R (1988), *Sensory Analysis of Food*, 2nd edn, Elsevier Applied Science, London.

PIGGOTT J R (1995), 'Design questions in sensory and consumer science', *Food Qual Preference*, **6**(4), 217–20.

ROCHA S, DELGADILLO I, CORREIA A J F, BARROS A and WELLS, P (1998), 'Application of an electronic aroma sensing system to cork stopper quality control', *J Agric Food Chem*, **46**(1), 141–51.

ROSENTHAL A (1999), *Food Texture, Perception and Measurement*, Aspen Publishers, Gaithersburg, Maryland.

SCHALLER E, BOSSET J O and ESCHER F (1998), '"Electronic noses" and their application to food', *Lebens-Wiss u-Technol*, **31**, 305–16.

SHAMIL S H, WYETH L J and KILCAST D (1992), 'Flavour release and perception in reduced-fat foods', *Food Qual Preference*, **3**(1), 51–60.

SMITH G L (1988), 'Statistical analysis of sensory data', in *Sensory Analysis of Foods*, 2nd edn, J R Piggott (ed), Elsevier Applied Science, London, 335–79.

STONE H and SIDEL J L (1993), *Sensory Evaluation Practices*, Academic Press, Florida.

3
Instrumental methods in detecting taints and off-flavours

W. J. Reid, Leatherhead Food International, UK

3.1 Introduction

The instrumental analysis of taints and off-flavours is a complementary technique to that of sensory analysis and presents its own interesting blend of certainties and challenges, largely governed by the sensory characteristics of the compounds under study. The presence of a tainting compound in a sample is apparent from the change in odour or flavour. The analyst begins, therefore, by knowing there is something to find. The description of the taint provides an additional parameter, because any target compound identified in the sample must have the same taste and odour characteristics as those derived from sensory analysis. Many of the compounds that cause taint do so because they are very potent and can be perceived at extremely low concentrations by human senses. For example, 2,6-dibromophenol has a taste threshold in water of $5 \times 10^{-4} \mu g\, l^{-1}$ (Whitfield et al., 1988). This presents a challenge to the analyst, given the need to detect these very low concentrations of the tainting compound, often in the presence of much higher concentrations of other naturally occurring materials from the sample. This threshold data is also a criterion for the analysis, since any candidate compound identified by the chemical analysis must be present in the sample at a high enough concentration to cause the perceived taint.

Generally, taint analysis falls into one of two categories. In one case, the compounds causing the taint are known and the analysis can be optimised for those compounds and the sample matrix. In the other case, the nature of the tainting compounds is not known and a more general approach is needed, often using sensory data in tandem with the instrumental analysis. The aim of this chapter is to examine the various methods of analysis used for tainting compounds in food. It will start with solvent-based extraction techniques, for example, liquid–liquid extraction, steam distillation or combined steam distillation and solvent extraction. The next section will deal with headspace methods such as static and dynamic headspace sampling and closed loop stripping. This will be followed by a description of the relatively recent technique of solid phase microextraction (SPME) and some applications of the technique to taint analysis. The next section will consider instrumental analysis, including combined gas chromatography–olfactometry (GCO), gas chromatography with selective detectors and gas chromatography–mass spectrometry (GCMS), while the final section will cover new methods in taint analysis such electronic noses and stir-bar sorbtive extraction (SBSE).

3.2 Liquid-based extraction techniques

The methods used for the extraction and concentration of tainting compounds are similar to those used for the extraction of other organic compounds and, in particular, those used for flavour compounds (Wilkes *et al.*, 2000). The difference lies in the need to detect the very low concentrations of the target compounds that can cause taint in the sample. The aim of the extraction is to arrive at a solution of the tainting compounds in a form that can be introduced into the analytical instrument. It is often useful to work back from the requirements of the instrument in order to define the requirements of the extraction.

Consider the situation where a specific compound is to be analysed, using a detection system such as high resolution gas chromatography (HRGC) coupled to a sensitive detector such as a mass spectrometer (MS) working in the selected ion monitoring mode (SIM) or, for halogenated compounds, an electron capture detector (ECD). It is reasonable to suppose that a peak with a good signal-to-noise ratio could be obtained from 1 pg of material injected into the instrument. If the final volume of the solution obtained from the extraction was 0.2 ml and the injection volume was 1 µl, then the extract will need to contain around 200 pg of material. It follows, therefore, that to detect 2,4,6-trichloroanisole at its taste threshold of $0.01 \mu g\,l^{-1}$ in wine (Buser *et al.*, 1982) it would be necessary to extract 20 ml of wine. If we aim for a detection limit at a concentration of one hundredth of the taste threshold, then it would be necessary to extract 2000 ml of wine. On the other hand, setting a detection limit for 2-bromophenol at one hundredth of its taste threshold in prawns, which is $2 \mu g\,kg^{-1}$ (Whitfield *et al.*, 1988), would require the extraction of 10 g of prawns. In cases where the aim is to identify an unknown compound it may be necessary to extract greater amounts of sample. It is apparent that any extraction procedure requires the removal of material that might interfere with the analysis, particularly given the extreme concentration factor required.

3.2.1 Solvent extraction

The simplest method of extracting organic compounds from any matrix is solvent extraction. In some cases the extract can be used directly for analysis. Indole and skatole have been associated with a taint in the flesh of intact male pigs. A high-performance liquid chromatographic (HPLC) method for these compounds in pig fat and meat has been developed in which a simple solvent extract is applied directly to the HPLC system (Garcia-Reguero and Ruis, 1998; Ruis and Garcia-Reguero, 2001), using a fluorescence detector which is selective enough not to respond to any co-eluting material. More usually, however, the extraction stage is followed by a series of clean-up stages to prepare the extract for analysis, often using the sensory characteristics of the tainting compounds to guide the procedure. Whitfield and co-workers (Whitfield *et al.*, 1991) used Soxhlet extraction with pentane to examine an earthy taint in flour. They extracted 250g of flour, concentrated the pentane extract and subjected it to simultaneous steam distillation and extraction (SDE) to recover the volatile compounds. The GCMS analysis was used to identify the tainting compound as geosmin.

An investigation into a medicinal taint in cake mix (Sevenants and Sanders, 1984) began with a solvent extract of the material. The extract was then partitioned between different solvents and the fraction containing the medicinal taint was separated into basic, neutral and acidic fractions. The acidic portion, which had the medicinal smell, was analysed by HRGC using an olfactometer and a flame ionisation detector (FID) in parallel. The extract was re-analysed by GCMS in an attempt to obtain a spectrum at the retention time identified from the olfactometry measurements. In this case the GCMS data proved to be complex around the region of interest. The analysis was simplified by the discovery that a mixture of minor ingredients of the cake mix produced the same taint. Extraction, clean-up and analysis of this mixture resulted in the identification of 2-iodo-4-methylphenol, which was shown to be formed by the reaction between cresol in flavouring and iodine from sea salt, two of the components of the mixture. A similar procedure was used to identify 2,6-dibromophenol in prawns (Whitfield *et al.*, 1988). The prawns were macerated in water and the mixture extracted in a liquid–liquid extractor in order to produce enough material for the identification of the compound. Subsequent target analyses used less material and relied on SDE for the sample preparation.

3.2.2 Supercritical fluid extraction

Under elevated pressures and temperatures, compounds that are normally gases reach a supercritical state in which they possess properties characteristic of both gases and liquids. In particular they have the penetrative power of a gas and the solvating power of a liquid. Supercritical fluid extraction (SFE) is widely used in food analysis and has been extensively reviewed (Anklam *et al.*, 1998; Chester *et al.*, 1998; Camel, 1998; Sihvonen *et al.*, 1999). In a recent book, Mukhopadhyay (2000) provides a detailed discussion on the theoretical aspects of SFE.

The fat of intact male pigs can have an objectionable odour. Androstenone and skatole have been reported to contribute to this taint. In an effort to eliminate the use of the solvents required for a normal liquid extraction, SFE was studied as an extraction method for these compounds (Zablotsky *et al.*, 1993; Zablotsky *et al.*, 1995). The authors used a static extraction system and found that they could achieve a recovery of around 97% for androstenone and around 65% for skatole. Although the recovery of skatole was low, it was found to be reproducible. A dynamic extraction system was also used to extract androstenone from pig fat (Magard *et al.*, 1995). Here the supercritical fluid passed through the sample and was led out from the apparatus through a trap containing octadecyl silica. After extraction the trap was eluted with 1.5ml of dichloromethane to recover the compound. The various parameters governing the extraction were studied in order to maximise the recovery of androstenone and minimise the extraction of fat. It was found that the solubility of the fat increased with temperature and pressure and so the lowest possible temperature and pressure, consistent with good recovery of the compound, were used. The analysis using SFE was compared with those from two standard methods based on radioimmunoassay and HPLC and the results from the three methods were found to be comparable.

A perennial problem for wine producers is the 'cork taint' in wine, which is reported to be the result of contamination by 2,4,6-trichloroanisole (Buser *et al.*, 1982) which originates in cork stoppers. A dynamic SFE-based method was developed for the analysis of 2,4,6-trichloroanisole in cork (Taylor *et al.*, 2000) with the aim of eliminating the solvent used in Soxhlet extraction and reducing the time of the analysis. After extraction the compound was eluted from the trap in methanol and analysed by GCMS. The results from SFE were compared with those from Soxhlet extraction using methanol and found to be comparable in both sensitivity and reproducibility. The SFE method was also rapid and almost solvent free.

3.2.3 Steam distillation

Most of the compounds that cause taint or off-flavour can be assigned to the class of volatile flavour components. The only difference is that the taste of the compound is unpleasant in the particular food in which it is found. Some compounds, such as 2,4,6-trichloroanisole, which gives a 'musty' taint, are unpleasant in any food. Other compounds, such as guaiacol, are regarded as having a positive impact in smoked food or in some whiskies but are seen as taints in vanilla ice cream or apple juice. Steam distillation has been used for the extraction of flavour and odour compounds for some considerable time. It relies on the volatility of those compounds, which co-distil with the steam, separating them from the non-volatile components, which remain in the sample. The distillate can be treated in various ways. For example, in the analysis of guaiacol in wine (Simpson et al., 1986), the distillate was collected in a basic solution to retain the analyte. It was then acidified and extracted with dichloromethane, which was concentrated to a low volume for GCMS analysis. In an investigation into a taint in melons (Sanchez Saez et al., 1991) the distillate was collected in dichloromethane in an ice bath. The dichloromethane was then used to extract the tainting material from the collected water. Analysis identified the compound responsible as 4-bromo-2-chlorophenol.

In order to avoid overheating the sample, which might lead to the breakdown of thermally unstable compounds, distillation can be carried out at lower temperature by placing the sample under vacuum. In an investigation of a phenolic taint in cheese, the sample was shredded and the curd removed (Mills et al., 1997). The remaining water and fat mixture was steam distilled, under vacuum, at a temperature of 60°C. The distillate was collected in a cold trap and extracted with diethyl ether. The extract was analysed by combined HRGC–olfactometry (GCO) and GCMS. 2-Bromo-4-methylphenol was identified as the probable cause of the taint. A similar method has been used to study geosmin in the Nile tilapia fish (Yamprayoon and Noohorm, 2000a; 2000b). Vacuum distillation can also been used to prepare large quantities of extract for analysis, for example, in the case of Mexican coffee which was found to have a mouldy or earthy taint (Cantergiani et al., 2001). A 100g sample was vacuum distilled and around 200–300ml of distillate was collected. This process was repeated five times and the distillates were pooled and extracted with dichloromethane, which was then concentrated to 0.5ml for analysis. The extract was examined by GCO and fractions isolated by preparative HRGC. The fractions were analysed by GCMS and a number of compounds were identified, including geosmin, methylisoborneol and 2,4,6-trichloroanisole.

A recent development of the technique has been the use of microwave-assisted steam distillation. In 1996 Conte and co-workers described a system for the extraction of geosmin and methylisoborneol from catfish (Conte et al., 1996a). Later that year the authors described improvements to the apparatus (Conte et al., 1996b). The sample was placed in a microwave oven, in a vessel linked to a C18 silica solid phase adsorbent cartridge. Internal standards of cis-decahydro-1-naphthol and endo-norborneol acetate were added and a stream of argon was passed over the tissue and through the cartridge while the microwave was operated at 40% power for 10min. The volatile organic compounds were then eluted from the trap with ethyl acetate and analysed by GCMS, with detection limits in the sub-parts-per-billion range (Conte et al., 1996b). The use of the solid phase trapping reduced the solvent required in the analysis to 2ml and the use of a microwave oven cut the extraction time to 10min. The method was further refined by the addition of SPME extraction (Zhu et al., 1999). The distillate from the sample was collected in a vial and the target compounds were extracted by SPME and injected into a GCMS for analysis. This completely eliminated the need for solvent and simplified the final stage of the extraction, making it more rapid to use. The authors investigated the recoveries of analytes using both methods. They concluded that the original solid phase trapping produced better recoveries, possibly because it was a closed system where all of the material distilling from the sample passed through the trap. In the other method, collection of the distillate took place in an open system, with the potential for evaporative loss of some of the distillate and the target compounds. The authors concluded, however, that use of SPME did not result in any loss of sensitivity or reproducibility.

A similar apparatus was used for the analysis of geosmin and methylisoborneol in catfish (Lloyd and Grimm, 1999). Deuterated analogues of the target compounds were added to fish tissue and the volatiles were distilled out using microwave-assisted distillation with nitrogen as a purge gas. Again SPME was used to extract the analytes from the distillate for GCMS analysis. The authors studied the effects of a number of parameters, including the operating power of the oven and the flow of the purge case, in order to optimise the extraction. They also compared the procedure with microwave distillation–solvent extraction and purge and trap–solvent extraction. They concluded that the SPME extract of the distillate produced the largest response from the GCMS analysis and that it gave good reproducibility. They reported a disadvantage of SPME in that it was only practical to carry out a single GCMS analysis from the distillate. With solvent extraction, repeat injections were possible. Quantitative analysis of geosmin and methylisoborneol at the low part per trillion range was possible using this method (Grimm et al., 2000a). Grimm and co-workers also carried out analysis of catfish using a version of the apparatus in which no purge gas was used (Grimm et al., 2000b).

3.2.4 Combined steam distillation and solvent extraction

One of the most popular methods for the extraction of volatile tainting compounds is SDE, usually using a variation of the apparatus first described by Likens and Nickerson (Likens and Nickerson, 1964). A review (Chaintreau, 2001) describes the original apparatus and some of the changes and improvements that have been made to it. The principle of operation of the Likens–Nickerson apparatus is straightforward. Water containing the sample is boiled in one flask while the extracting solvent is boiled in another. The vapours are allowed to mix and condense in a central chamber. The immiscible condensates form two layers at the bottom of the central chamber and return pipes lead the water and solvent back to their respective flasks. Volatile compounds distil out of the sample with the steam and are extracted into the solvent during the mixing and condensing process. As the solvent passes back into its flask it carries the extracted volatiles with it. Over a period of time the volatile compounds will be extracted into the solvent.

Because the volatilisation of the solvent followed by the condensation, extraction and return form a cyclic process, a minimal amount of solvent can be used. It is possible to use relatively large sample sizes, since it is only the steam and the entrained volatiles that enter the body of the extractor. In the author's laboratory, 100 g samples are regularly dispersed in 1 l of water and extracted using 30 ml of diethyl ether. After extraction the solvent can easily be concentrated to a low volume for analysis.

Good recoveries can be obtained with this equipment. Whitfield reported recoveries of 80–100% for 2,4,6-trichloroanisole, 2,3,4,6-tetrachloroanisole and pentachloroanisole (Whitfield *et al.*, 1986) and 93% recovery of 2,4,6-tribromoanisole (Whitfield *et al.*, 1997) from both fruit and fibreboard. Recoveries do vary with the nature of the analyte. Steam distillation and extraction of coffee suspensions showed similar recoveries of the anisoles but lower values for chlorophenols. The figures were 70–76% for the 2,4,6-trichlorophenol, 35% for 2,3,4,6-tetrachlorophenol and 18% for pentachlorophenol (Spadone *et al.*, 1990). It is possible to improve the recoveries of basic or acidic compounds by changing the pH of the aqueous mixture. For example, in the quantitative analysis of 2,6-dibromophenol in crustacea the mixture of water and sample was acidified to pH 2 with sulfuric acid (Whitfield *et al.*, 1988) to force the compound into its free acid form and increase its volatility. Recoveries can be adversely affected by a high lipid content in the sample (Au-Yeung and MacLeod, 1980). To counter this problem, a modification of the Likens–Nickerson apparatus used a steam generator to inject steam into the sample during extraction. This improved the recoveries of each of the compounds studied (Au-Yeung and MacLeod, 1980).

A potential problem with SDE is the formation of artefacts during the extraction procedure. A comparison of SDE and vacuum distillation of hen meat (Siegmund *et al.*, 1997) showed that thialdine could be detected in the extract from SDE but not in the extract from vacuum distillation. Investigation of the process showed that thialdine was being produced during the SDE extraction. One approach to reducing artefact formation has been the use of a Likens–Nickerson extractor modified to operate at reduced pressure (Chaintreau, 2001). Vacuum SDE was used to extract geosmin from beans, with the aqueous mixture boiling at 45°C (Buttery *et al.*, 1976). Another version of a vacuum Likens–Nickerson apparatus was also described by Schultz and co-workers (Schultz *et al.*, 1977). Operation under vacuum requires the use of a relatively non-volatile extracting solvent to avoid losses during the extraction (Chaintreau, 2001).

3.3 Headspace extraction

Solvent-based extractions are not suitable for the analysis of very volatile compounds. The compounds can be lost during the extraction and clean-up stages and those that are present in the final extract can be impossible to detect as they are masked by the presence of the extracting solvent. An alternative approach to analysis of these compounds is to use their volatility as part of the separation step, collecting them from the headspace of the sample. Headspace techniques have the advantage that they can be carried out at relatively low temperatures, minimising the possibility of artefact formation and they require little or no solvent, depending on the method in use.

3.3.1 Static headspace sampling

Static headspace sampling is the simplest of the available techniques. A portion of the sample is placed in a sealed container and allowed to stand until the concentration of volatile compounds in the headspace above the sample has reached equilibrium. The sample can be heated to increase the concentration of analytes in the headspace. A volume of the headspace is then removed and introduced into the analytical instrument. The sensitivity of this approach is governed by the volume of headspace taken from the sample and the concentration of the target compounds in the headspace at the time of sampling.

Sorbic acid, which is used as a preservative, can be decarboxylated by some species of *Penicillium* mould to produce 1,3-pentadiene which has an odour described as 'paint', 'kerosene' or 'plastic' (Marth et al., 1966). Static headspace has been used for the analysis of 1,3-pentadiene in cheese spread (Daley et al., 1986), margarine and cheese (Sensidoni et al., 1994). It was also used, in conjunction with GCMS and combined gas chromatography–infrared spectroscopy (GCIR), in the investigation of a musty taint in packaging film (McGorin et al., 1987). The simplicity of static headspace sampling lends itself to automation and several commercial headspace injectors are available for HRGC instruments. In an investigation of autoxidation in wheat germ, hexanal was chosen as a marker compound for the development of off-flavour and was measured in a number of samples using a headspace autosampler.

3.3.2 Dynamic headspace sampling

In order to overcome some of the sensitivity limitations of static headspace sampling, a dynamic headspace technique, also known as purge and trap sampling, has been used. The sample is placed in a container fitted with a purge head connected to a trap. A clean inert gas is passed over a solid sample or through a liquid sample and allowed to exit through the trap. Volatile compounds are swept out of the headspace into the trap. Since the concentration of the analyte in the headspace is being depleted, the equilibrium is disturbed and more of the material will be transferred from the sample to the headspace and so, as the sampling is continued, the amount of the analyte held on the trap will increase. As with static headspace, heating the sample can increase the concentration of the compounds in the headspace and so increase the amount of sample on the trap.

For any particular compound the limiting factor of the trapping process is the breakthrough volume. As the purge gas passes through the trap it will act in the same way as the carrier gas in a gas chromatographic column. At some point the analyte will reach the end of the trap and begin to be eluted. From then onwards there will be a loss of material from the trap. The breakthrough volume will depend on the analyte, the material used in the trap, the purge gas and the temperature. The use of several different trapping materials has been reported in the literature. Horwood and co-workers used Poropak Q for the determination of 1,3-pentadiene in cheese (Horwood et al., 1981), while Chromosorb 105 was used for the analysis of trimethylarsine in prawns (Whitfield et al., 1983). A widely used trapping medium is Tenax, which is available in several forms. Tenax GC was used to analyse for the ethyl esters which were causing a 'fruity' off-flavour in milk (Wellnitz-Ruen et al., 1982), for styrene in apple juice (Durst and Laperle, 1990) and in the investigation of a musty taint in French fries (Mazza and Pietzak, 1990). Tenax TA was used for the identification of geosmin in clams (Hsieh et al., 1988) and in cultured microorganisms (Börjesson et al., 1993) and in the investigation of off-flavours generated in olive oil during oxidation (Morales et al., 1997). Tenax TA was compared with three carbon-based trapping materials for the analysis of off-flavours generated in whey protein concentrates during storage (Laye et al., 1995). It was concluded that Tenax TA was the best material for some of the compounds under study, while Vocarb 3000 was best for the rest of the compounds.

Generally, after the sampling phase is complete, the trapped volatiles are introduced into a HRGC column using some form of thermal desorption and cold-trapping system. The trap, containing the volatiles, is placed in an oven connected to a cold trap, which is connected in turn to the head of the HRGC column. Carrier gas passes through the adsorbent, which is heated to release the volatiles, which are then concentrated in the cold trap. After a suitable purge time, the cold trap is rapidly heated to transfer the volatiles to the HRGC column in a narrow band and the analysis is started. If the original sample contained water there is a possibility that moisture would be retained in the trap during the sampling process. This can lead to practical difficulties during the thermal desorption and cold-trapping injection. The water can be eliminated by purging the trap with clean dry gas after the sampling period is complete (Ramstad and Walker, 1992). The breakthrough volume for water in most trapping materials is considerably lower than those for the volatiles of interest and so water can be removed while retaining the analytes. Solvent extraction has also been used to remove the volatiles from the trap (Harris et al., 1986; Mazza and Pietzak, 1990). This has the advantage that more than one analysis of the trapped compounds can be undertaken but the corresponding disadvantage that the sensitivity of the analysis will be lower, since a fraction of the total sample is being analysed. Solvent extraction can only be used for the analysis of those compounds that are well separated from the solvent during analysis. Volatile compounds can be analysed directly from non-aqueous samples using thermal desorption and cold trapping without a preliminary sampling stage. In an analysis of geosmin in fish, oil extracted from the fish was placed onto a plug of glass wool in the injection port of a gas chromatograph, while the head of the column was cooled to $-30°C$. After the volatiles had condensed on the head of the column, the temperature was raised and the analysis started (Dupuy et al., 1986).

A variation on dynamic headspace sampling is the closed-loop stripping apparatus (CLSA) first described by Grob and co-workers (Grob, 1973; Grob and Zücher, 1976). Here the sample and the trap are part of a loop, which also contains a pump. The gas in the loop, either air or an inert gas, is continuously pumped through the system. Volatiles are purged from the sample and concentrated in the trap, which contains a small amount of a carbon-based adsorbent. As with an open system the limiting step will be the breakthrough volume of the analytes. Since this is a cyclic system, compounds eluting from the end of the trap will not be lost but will be carried back through the system in the gas stream. Thus the concentration of the compounds on the trap will reach equilibrium. After sampling is complete, the volatiles can be removed from the trap in a very small amount of solvent. It is possible, therfore, to analyse the extract of each sample more than once, with the disadvantage that the sensitivity is correspondingly reduced compared to a single-shot technique such as Tenax trapping. The extraction efficiencies of different solvents for a wide range of volatiles have been investigated (Borén et al., 1985). Closed-loop stripping apparatus has been used in the investigation of musty taints in water (Krasner et al., 1983; Martin et al., 1988) and in beet sugar (Marsili et al., 1994). It was used for the analysis of geosmin and 2,4,6-trichloroanisole in a study on the effectiveness of biological treatment in removing odorous compounds from water (Huck et al., 1995). Objectionable odours from the Lake of Galilee were identified using CLSA and GCMS and found to be caused by organosulfur compounds such as dimethyl sulfide, dimethyl disulfide, dimethyl trisulfide and methane thiosulfonate (Ginzburg et al., 1998).

3.4 Solid phase microextraction (SPME)

Since its introduction by Pauliszyn and co-workers (Arthur and Pauliszyn, 1990; Arthur et al., 1992; Potter and Pauliszyn, 1992) SPME has become widely used as an extraction technique. The principles of its operation are simple. A fused silica fibre, coated with one of a number of polymers, is introduced into the headspace above a sample or, for liquids, into the sample itself. Compounds from the material are adsorbed onto the fibre until equilibrium is reached between the adsorbent and the matrix. The fibre is then removed from the sample and introduced into the injection port of a gas chromatograph where the trapped compounds are thermally desorbed into the head of the column for analysis. A review covers the theory of operation in some detail (Pauliszyn, 2000). The equilibrium between the sample and the fibre is influenced by a number of factors and so the choice of fibre and sampling conditions can affect the recovery of the analytes and therefore the sensitivity of the method. A second review (Shirey, 2000) describes the optimisation of these parameters.

The SPME technique has a number of advantages. It is solvent-free, extraction is both simple and rapid and injection into the gas chromatograph requires no specialised equipment. The operations involved in sampling and injection are very similar to those of a static headspace injection. As a result the technique has proved amenable to automation and commercial systems are available which will carry out unattended extraction and analysis. A comparison between SPME and dynamic headspace analysis using Tenax, concluded that with a polydimethylsiloxane (PDMS) fibre, the two methods recovered similar compounds with comparable reproducibilities but that the Tenax analysis was more sensitive (Elmore *et al.*, 1997). On the other hand, in a study of light-induced lipid oxidation products in milk, it was concluded that SPME was the better technique and that it was cheaper to operate in practice (Marsili, 1999). Three different types of fibre were compared in the development of a method to analyse hexanal and pentanal as markers for the development of off-flavours in cooked turkey. The phases used were Carboxen/PDMS, PDMS/divinylbenzene (DVB) and DVB/Carbowax. For these compounds, the PDMS/DVB fibre showed the best combination of reproducibility, sensitivity and linearity (Brunton *et al.*, 2000). Derivatisation of the compounds trapped on a fibre can be carried out before GCMS analysis. Diazomethane has been used to methylate chlorophenols and other acidic compounds by exposing the fibre, after sampling, to the reagent in a gaseous form (Lee *et al.*, 1998).

An area where SPME is widely used is in the analysis of 2,4,6-trichloroanisole in wines and corks. Evans and co-workers developed a SPME method for the compound in wine, which was comparable in performance to existing methods but was much more rapid (Evans *et al.*, 1997). During the development of a method for the analysis of the compound in both wine and cork (Fischer and Fischer, 1997), the fibre was used to sample the headspace above either wine or ground cork. A comparison was made between sampling by immersion into the wine and sampling in the headspace. It was found that a better recovery of 2,4,6-trichloroanisole was obtained from the headspace. It was also suggested that dipping the fibre into the wine might lead to the transfer of non-volatile material to the injector of the gas chromatograph resulting in loss of performance. An automated analysis, using SPME in combination with GCMS, has also been described (Butzke *et al.*, 1998). In all of these methods the detection limit for 2,4,6-trichloroanisole in wine was around 5 parts per trillion (ppt) (ng l^{-1}).

The SPME technique has been used for the analysis of taints in a wide variety of materials. An off-flavour in margarine was identified as a series of ketones produced by a mould which had grown in the product (Hocking *et al.*, 1998). An off-flavour in milk, described as fruity and rancid, was shown to be due to a mixture of ethyl esters and fatty acids again formed by a microorganism which had contaminated the milk (Whitfield *et al.*, 2000). The flavours and off-flavours formed during storage of strawberry juice were determined by headspace SPME (Golaszewski *et al.*, 1998) and the compounds responsible for the off-flavours were identified. Investigation of an off-flavour in a dressing showed the origin to be in one of the ingredients, beef plasma protein (Koga *et al.*, 2001). The volatiles from the protein were analysed using SPME and GCMS. The compounds identified included hexanal, pentanal and 2-methylbutanal which are associated with oxidative rancidity. The SPME technique has been used to quantify acetaldehyde in spring water and various types of milk in a study to determine the flavour threshold of the compound in these products (Van Aardt *et al.*, 2001). Geosmin and 2-methylisoborneol have been determined in water using SPME and GCMS (McCallum *et al.*, 1998; Lloyd *et al.*, 1998). The technique has also been used in conjunction with microwave distillation for the analysis of geosmin and 2-methylisoborneol in fish. The distillate from the fish tissue was analysed using SPME rather than solvent extraction (Zhu *et al.*, 1999; Grimm *et al.*, 2000a; Grimm *et al.*, 2000b).

3.5 Gas chromatography and other methods

3.5.1 Gas chromatography–olfactometry (GCO)

Most investigations of taint or off-flavour rely on gas chromatography. There are numerous reviews and books discussing the fundamentals of the technique and its application to food chemistry (Siouffi and Delaurent, 1996; Le Bizec, 2000; Handley and Alard, 2001). When the tainting compounds are known, it is possible to use a selective detector, such as ECD or a mass spectrometer operating in SIM mode, to increase the specificity and sensitivity of the analysis. When the causes of the taint are not known they may be identified using GCMS, with the mass spectrometer operating in scan mode to collect spectra of the compounds eluting from the column. Sometimes it is possible to compare the total ion chromatogram (TIC) from the GCMS analysis of a tainted sample with that of an untainted sample and observe differences which correspond to the tainting compounds. More frequently, however, the responses for the taints are small and the differences are not obvious. In this case the investigation can be easier if the retention times of the tainting compounds can be found, allowing the analyst to concentrate on those areas of the TIC. One popular approach to identifying the retention characteristics is to use GCO, where the human nose is used as a selective and sensitive detector of the compounds of interest.

This technique has been widely used for some time in the analysis of flavour compounds, including tainting compounds, and the principles of the technique have been extensively reviewed (Acree, 1993; Blank, 1997; Feng and Acree, 1999; Friedrich and Acree, 2000). The basic idea of GCO is simple. The effluent from the gas chromatographic column is mixed with air and water vapour and led to a point where a human assessor can perceive the aromas of compounds eluting from the column. The entire effluent of the column can be used or a splitter can be inserted, to divert part of the effluent to the olfactometer and the rest to a detector, for example a mass spectrometer. There are a variety of parameters to be considered in the optimisation of the technique. For example, the flow of make-up air and the humidity need to be controlled to present the compounds to the observer in the most efficient manner (Hanaoko et al., 2000). The most significant factors, however, are those that affect the perception of aroma by the sensory panellist. These include the alertness of the panellist, the tiredness of the senses, olfactory adaptation and the ability of the assessor to detect particular compounds (Kleykers and Scifferstein, 1995; Friedrich and Acree, 2000; Van Ruth and O'Connor, 2001). Given the short elution time of a compound from the HRGC column, even the rate of breathing of the subject can affect the detection of the odour (Hanaoka et al., 2001).

Gas chromatography–olfactometry has also been used in the investigation of taints. For example, it was used by Patterson in 1970, during the investigations into taints in meat (Patterson, 1970). A selection of reports describes the use of GCO for the analysis of 2,4,6-trichloroanisole in coffee (Holscher et al., 1995), 2-bromo-4-methylphenol in cheese (Mills et al., 1997) and 2-aminoacetophenone in wine (Rapp, 1998). It has been used for the study of tainting compounds in cork (Moio et al., 1998) and for off-flavours in boiled potatoes (Petersen et al., 1999) and in soybean lecithins (Stephan and Steinhart, 1999). Darriet and co-workers described the use of GCO to identify geosmin as the cause of an earthy taint in wine (Darriet et al., 2000), while Escudero et al. (2000) used GCO to identify methional as the source of a vegetable off-flavour in wine.

3.5.2 Gas chromatography with selective detectors

In cases of taint or off-flavour when the identities of the tainting compounds are known and the molecules contain a suitable heteroatom, it is possible to carry out a gas chromatographic analysis using a selective detector. The detector can be chosen so that it will produce a response for the target compound but not for compounds which do not contain the heteroatom. This reduces the potential interference from any co-extracted compounds that elute at the same retention time as the tainting compounds. When the compound contains a halogen such as chlorine or bromine, it is likely to produce a signal on an ECD. The greater the number of halogens in the molecule the greater the response produced. An ECD has been used in the analysis of chlorophenols and chloroanisoles. In a study on Mason jars, pentachlorophenol was found to be present in the jar lids and to have migrated into various foods stored in the jars using home canning techniques (Heikes and Griffitt, 1980). The pentachlorophenol was extracted from the samples using dichloromethane at acid pH and methylated to form the pentachloroanisole, which was then determined by HRGC–ECD. Chlorophenols and chloroanisoles in wines and corks have been analysed by solvent extraction followed by HRGC–ECD (Bayonove and Leroy, 1994), as have chloroanisoles and chloroveratroles in water (Paasivirta and Koistinen, 1994), and chloroanisoles in pharmaceutical products (Ramstad and Walker, 1992).

Sulfur-containing compounds are present in many different foods and contribute to their flavour. For example, dimethyl sulfide contributes to the taste and flavour of beer (Scarlata and Ebeler, 1999). In some cases, however, an excess of one or more of the sulfur compounds can lead to an off-flavour. Analysis of sulfur compounds in various foodstuffs has been carried out using three types of element selective detector, the flame photometric detector (FPD), the sulfur chemiluminescence detector and the atomic emission detector. Mistry and co-workers compared these detectors and concluded that the atomic emission detector showed a greater sensitivity and a wider dynamic range than the other two (Mistry *et al.*, 1994). The atomic emission detector has been used to determine dimethyl sulfide and its precursors in red wine (Swan, 2000) and also for the analysis of chlorophenols, at parts per trillion levels, in drinking water (Turnes *et al.*, 1994).

Analysis of sulfur compounds produced during UHT (ultra high temperature) treatment of milk (Steely, 1994) and of sulfur compounds in brandy (Nedjma and Maujean, 1995) has been carried out by HRGC combined with the chemiluminescence detector. A study of its use in wine examined the optimisation of the operating parameters and concluded that it was more sensitive for volatile sulfur compounds than the FPD (Lavigne-Delcroix *et al.*, 1996). A review on the analysis of sulfur compounds in wine describes the use of both chemiluminescence and the FPD (Mestres *et al.*, 2000). The FPD detector has been used by Mestres and co-workers to analyse volatile sulfur compounds in wine using static headspace extraction (Mestres *et al.*, 1997) and headspace SPME (Mestres *et al.*, 1998; 1999a; 1999b). They concluded that a commercially available fibre, Carboxen/PDMS, was the most suitable for this analysis (Mestres *et al.*, 1999b). The combination of headspace SPME and chromatography with an FPD has also been used to analyse beer for dimethyl sulfide (Scarlata and Ebeler, 1999) and a number of other volatile sulfur compounds (Hill and Smith, 2000). The FPD has also been used in the analysis of dimethyl sulfide in rice and its products (Ren *et al.*, 2001) as well as in heat-treated milk (Bosset *et al.*, 1996).

3.5.3 Gas chromatography–mass spectrometry

The combination of gas chromatography and mass spectrometry has resulted in the most powerful tool available for the analysis of volatile organic compounds. There are many books available both on the subject of mass spectrometry and GCMS. Examples include those by De Hoffmann and Stroobant (2001) and by Hubschmann (2001).

When a MS is used in conjunction with a gas chromatograph there are two common modes of operation, full scan and SIM. In full scan analysis the mass spectrometer is set to scan continually over a given mass range and record complete mass spectra. If the total signal in each spectrum is summed and these summed values are plotted against time, the result is a TIC, which resembles the output from a non-specific GC detector such as an FID. The mass spectrum at any point in the trace can be examined and searched against existing libraries of spectra that are usually included with the data system of any GCMS instrument. The spectrum of an unknown can also provide information on the structure of that compound. Full scan mass spectrometry is often used to carry out a survey analysis, identifying the volatile compounds present in a sample. In situations where enough material is present in the extract, full scan GCMS can also give a very selective quantitative assay for a target compound, since the combination of mass spectrum and retention time gives a high degree of confidence in the identification.

Selected ion monitoring mass spectrometry is used for the identification and quantitation of known compounds. The MS is set up to record data from a small number of ions, characteristic of the target compounds, rather than scanning the entire mass range. The result is an improvement in sensitivity of the analysis, with a minimal loss in the selectivity. In SIM analysis, the identification of a compound is based on the elution of the target ions at the correct retention time. Additionally the ratios of the peak areas due to each ion should be the same in the target compound as in a reference standard. A good example of the use of both full scan and SIM was in the work carried out on 2,6-dibromophenol in prawns (Whitfield *et al.*, 1988). The initial analyses used full scan GCMS to identify the tainting compound. Subsequent quantitative analyses used SIM, allowing a much simpler sample preparation to be used, because of the improved sensitivity of the analytical method.

GCMS analysis allows the use of compounds containing stable isotopes as internal standards. For example, some or all of the hydrogen atoms in the molecule can be replaced with deuterium; alternatively, ^{13}C can be used to replace some or all of the ^{12}C atoms. The result is a stable molecule that has chemical and physical properties very similar to that of the original

46 Taints and off-flavours in food

compound. The labelled molecule is very unlikely to occur naturally, so there is little risk of interference from molecules of the internal standard naturally present in a sample. Additionally, the mass spectrum of the stable isotope shows the same breakdown pattern as that of the original molecule but the fragments have different masses. The shift in the mass will depend on the number of altered atoms in the fragment. In a SIM analysis the same fragments can be monitored in the target compound and the standard by setting up two masses for each fragment. The disadvantage to the use of any internal standard is that it does reduce the time that the mass spectrometer spends detecting each ion, since more ions are being detected in the same time; this can lead to a drop in sensitivity.

Deuterated toluene has been used as an internal standard for the analysis of a range of volatile compounds by purge and trap GCMS (Zhou et al., 2000), making use of the fact that the compound is unlikely to occur naturally. Stable isotopes have been used in the analysis of 2-aminoacetophenone in wine (Dollman et al., 1996), sulfur-containing compounds in wine (Kotseridis et al., 2000) and 2,4,6-trichloroanisole in cork stoppers (Taylor et al., 2000).

The analysis of 2,4,6-trichloroanisole in wine has also been carried out using an isomer of the compound as an internal standard (Aung et al., 1996). Here, 2,4,5-trichloranisole was used. This compound is not known to occur naturally, so there is little risk of interference from the samples. The internal standard has the same fragment masses as the target compound but a different chromatographic retention time. Both compounds can be detected in SIM, using the same target ions, without any reduction in the sensitivity of the analysis.

Some mass spectrometers can be used to distinguish between masses that differ by small amounts, which allows them to make use of the differences in masses that are due to the elements present in a fragment. The masses of the atoms of each element are not whole numbers, except for that of the most prevalent isotope of carbon, which has been set to 12 and from which all other elements are measured. For example, the mass of hydrogen is 1.00794, so compounds containing only carbon and hydrogen have masses that are greater than integers, i.e. mass sufficient. Some elements, such as the halogens, have masses that fall just below the nearest integer, so a compound containing enough of these elements will have a mass just below the integer value, i.e. it will be mass deficient. For example, tetradecene has a mass of 196.21910, while trichlorophenol has a mass of 195.92495.

These properties can be used to detect halogenated compounds in a sample. If full scan GCMS is carried out with a mass spectrometer capable of distinguishing between the mass deficient and mass sufficient fragments, the data from the analysis can be inspected to determine if any mass deficient species are present. The data systems produced by some instrument manufacturers allow this to be done automatically by applying filters to the data from the analysis. For example, a total ion chromatogram or a mass spectrum could be generated using only ions with fractional masses between 0.7 and 0.99. This would screen out all the mass sufficient ions in the analysis, leaving data only from potentially halogenated compounds. These techniques have been used in the author's laboratory to identify brominated and iodinated phenols and cresols.

An alternative approach to the identification of halogenated species is the use of negative ion chemical ionisation (NCI). Under these conditions many halogenated species produce a fragment consisting of a negatively charged halogen atom. It is possible to use reconstructed ion chromatograms of the halogen ions to examine scan data or to carry out the analysis using the ions in SIM mode (Sanders and Sevenants, 1994). The combination of these techniques has also been successfully used in the author's laboratory to identify brominated and iodinated compounds. Chemical ionisation can also be carried out in the positive mode (PCI). A combination of electron impact and positive chemical ionisation has been used in the analysis of geosmin and 2-methylisoborneol in water (McCallum et al., 1998).

3.5.4 High-performance liquid chromatography (HPLC)
Although most tainting compounds are amenable to HRGC-based methods, there are some cases where HPLC is a more suitable method. For example, in the analysis of indole and skatole in pigs, HPLC has been used as the basis of a simple and rapid assay (Tuomola et al., 1996; Garcia-Regueiro and Ruis, 1998; Ruis and Garcia-Regueiro, 2001). It has also been used to study the formation of 4-vinylguaiacol in orange juice (Lee and Nagy, 1990).

In cases where the tainting compounds are large or polar molecules, HPLC is the only method available for their analysis. Off-flavours in milk, produced by the Maillard reaction between lactose and lysines, have been studied using HPLC with a diode array detector and HPLC coupled to a mass spectrometer (LC-MS). Tentative identifications of some of the products were possible (Monti et al., 1999). Development of off-flavours in fatty foods can be the result of oxidation of the lipids to volatile compounds such as aldehydes and ketones. This reaction proceeds by the formation of hydroperoxides which then break down to produce the smaller compounds. The formation of the hydroperoxides has been studied using methods based on HPLC (Akasaka et al., 1999).

3.6 Stir-bar sorptive extraction

Stir-bar sorptive extraction (SBSE) is a technique based on the principles of SPME. The sampler is a small stirrer with a bar magnet encased in glass.

48   Taints and off-flavours in food

The surface of the glass is coated with a polymer in a similar way to the surface of a SPME fibre. As the bar is used to stir an aqueous sample, organic compounds partition between the sample and the polymer. After sampling is complete the bar is removed, washed with clean water and carefully dried. The bar is then inserted into a thermal desorption and cold trapping injector which transfers the trapped compounds onto the head of a GC column for analysis. The system has all the advantages of SPME but with a greater capacity for organic compounds, owing to the increase in the surface area of the adsorbent. It has been used to analyse geosmin, methylisoborneol and 2,4,6-trichloroanisole in drinking water (Ochiai *et al.*, 2001) with detection limits below 1 ppt. It has also been investigated as a tool for the determination of various types of taint in wine, packaging and breakfast cereal (Offen *et al.*, 2001).

3.7 Electronic noses

Electronic noses are a relatively recent development, with commercial applications of the technique starting to appear in the mid-1990s. A good introduction to the field is given in the book by Gardner and Bartlett (1999). Learning from the functional principles of the human nose, the electronic nose comprises an array of electronic chemical sensors with broadly overlapping specificities and an appropriate pattern recognition system capable of recognising simple or complex odours (Persaud and Dodd, 1982; Persaud and Travers, 1997).

3.7.1   Principles of electronic noses

The principle of operation of electronic noses is simple. Headspace gas from a sample is introduced to an array of sensors, which will respond to different degrees depending on the nature of the organic compounds present. The result is a histogram with each sensor represented by a vertical bar whose height is the intensity of response. In some ways, this resembles a mass spectrum and, in fact, mass spectrometers are the basis of some electronic noses. The data from the histogram of responses from different samples are then assessed statistically and differences between the samples can be plotted. Electronic noses can also be trained to detect certain types of odour pattern.

Current systems generally employ fewer than 50 sensors and are still far removed from a true 'electronic nose' with the full complexity of the human nose with its 10^7 to 10^8 receptors and part of the central nervous system for signal processing. However, the application-specific electronic nose (ASEN) with sensors and algorithms specialised for a specific application offers a cost-effective option for the on-line monitoring of flavours and off-flavours.

3.7.2 Components of an application-specific electronic nose system

Figure 3.1 shows a schematic diagram of the application-specific electronic nose (ASEN) system of odour detection. Components **1** to **3** of the system comprise the odour delivery system which provides control over the sample's presentation to the sensor array. The array of sensors (component **4**) comprises sensors of one or more transducer types as base devices. These carry the chemically sensitive layers selected for the application.

For aroma assessment, the sensors most frequently employed are polymeric chemoresistors. For applications such as the identification of

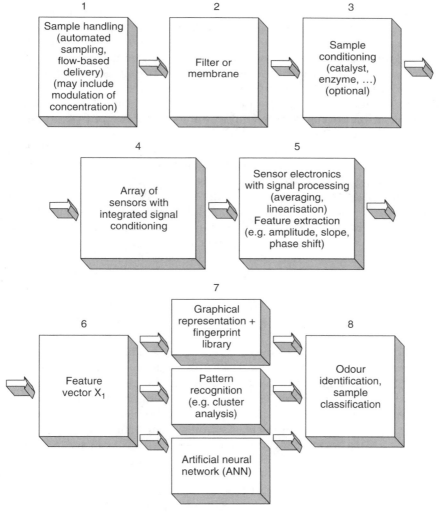

Fig. 3.1 Components of an application-specific electronic nose (ASEN) system.

off-odours, metal oxide (MeOx) semiconductors are employed more frequently. The traditional material here is tin oxide (SnO_2) which is the basis of many commercial sensors for industrial and domestic monitoring of combustible gases. A more recent material in commercial MeOx sensors is Ga_2O_3. Other semiconducting metal oxides have been examined for their gas sensing characteristics (Lampe *et al.*, 1996; Kohl, 1997; Capone *et al.*, 2000). The sensitivity characteristics of MeOx gas sensors are tuned by varying the preparation of the material (doping and sintering), the contact area (Hoefer *et al.*, 1997) and/or the operating temperature. They respond to gases and volatiles such as H_2, CH_4, NH_3, CO, NO_x, SO_x, H_2S and alcohols. Their molecular receptive range is more limited than that of polymeric chemoresistors. Less frequently used are the organic semiconductors such as the phthalocyanines (Huo *et al.*, 2000). An interesting approach is the use of the intact chemoreceptors of insects as odour sensors. Some arthropods have extraordinary sensory abilities and these can be harnessed by interfacing their chemoreceptive organs to microelectronic devices (Schütz *et al.*, 1996; 1997). Blowfly receptors coupled to microelectrodes detected 1,4-diaminobutane in the range 1 ppb to 100 ppm (decaying meat odour), butanoic acid between 20 ppm and 200 ppm and 1-hexanol from 8 ppm to 500 ppm (Huotari, 2000).

A combination of transducer types and sensing layer types (hybrid sensor systems) can be employed within an ASEN system. The molecular receptive range (MRR) can be widened and different aspects of the same odour molecule can be captured. As an example, functional groups can be probed with conducting polymers, molecular mass can be measured with piezoelectric devices, steric selectivity can be achieved with lipid layers or through functional side groups of polymers. In some odour-sensing applications, it can also be meaningful to include a specific chemical sensor such as a biosensor with an immobilised enzyme, an immunosensor with an immobilised antibody or a DNA sensor.

The result after processing by component **5** is an output in the form of the feature vector (component **6**). This feature vector is fed into the evaluation module (component **7**) where it can now be plotted in the chosen graphical representation such as bar chart (Fig. 3.2) or web chart (Fig. 3.3) and compared to a fingerprint library of odours. Additionally, feature vectors for a population of odour samples can be analysed using pattern recognition techniques (PARC). Each sampled odour can be described by its principal components and clusters of odour samples falling into a specified range for one or more principal components can be identified (Fig. 3.4). Alternatively, an artificial neural network (ANN) can be trained to interpret the pattern (Gardner and Hines, 1997). Neural networks can be combined with fuzzy logic (Berrie, 1997) into neuro-fuzzy systems (Theisen *et al.*, 1998). Either way, the ASEN arrives at component **8**, odour identification and classification of the sample in terms of flavour or off-flavour attributes.

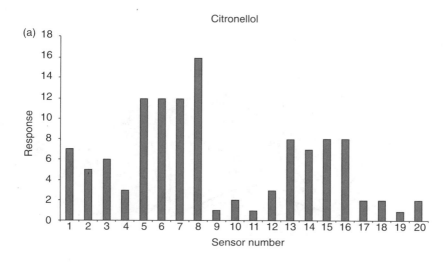

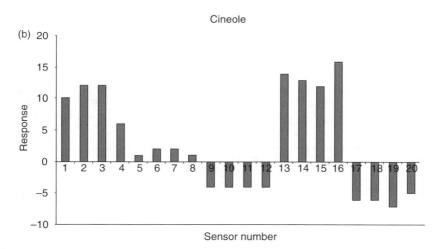

Fig. 3.2 Bar chart representation of the feature vectors for two aromas, (a) citronellol and (b) cineole (after Persaud and Travers, 1997).

A new addition to the spectrum of electronic nose technologies is the fingerprint mass spectra system (FMS). This instrument (Dittmann et al., 1988; Shiers et al., 1999; Dittmann and Nitz, 2000) is based on a quadrupole mass spectrometer combined with a headspace sampler and a computer. Volatile sample components are introduced into the MS without separation thus creating a mass spectrometric pattern. This reduced mass spectrometric pattern is analysed with the pattern recognition techniques developed

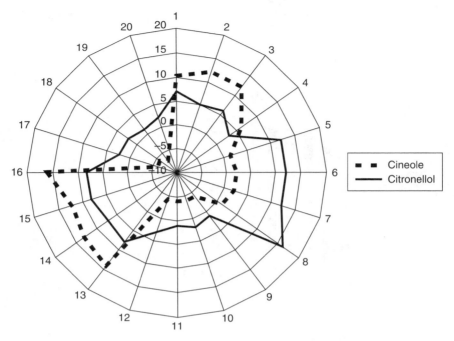

Fig. 3.3 Web chart representation of the vectors for the aromas citronellol and cineole.

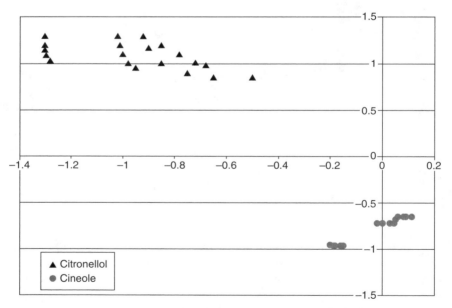

Fig. 3.4 Patterns generated by cluster analysis of the aromas citronellol and cineole (the first principal component is plotted along the horizontal axis, the second along the vertical axis).

for the electronic noses as described previously. A database of known reference samples is created and this can be used later to identify the property of interest in a new sample. The use of the label 'electronic nose' for the FMS systems has apparently prompted some manufacturers of electronic noses (as in the original definition) to rename their system as sensor arrays technology (SAT) (Mielle and Marquis, 2000; Mielle et al., 2000). Mielle and co-workers note that the competition between the manufacturers of these two types of systems – SAT and FMS – has been very aggressive since 1998. The respective advantages have been hotly debated. Supporters of the FMS systems emphasise their selectivity, adaptability and sensitivity. They concede that a well-equipped chemical analysis laboratory and some investment in time and effort would be required for the screening stage prior to training and final measurement. Supporters of the SAT systems argue that the sensitivity and accuracy are comparable and emphasise the smaller size and cost with the potential for further reductions in both size and cost. The issue of assay speed is a complex one. The sample throughput is related to variables such as sampling time, headspace equilibration time, recovery or cleaning time.

3.7.3 Uses of electronic noses

Evans et al. (2000) tested an electronic nose based on a conducting polymer array plus a neural network against a human sensory panel for the quality assessment of wheat with respect to the absence of taints. After training the system with 92 samples, a predictive success of 93% was achieved.

Electronic noses have been used as a rapid screening method for taints in packaging (Pitt, 1996; Shiers and Squibb, 1999; Forsgen et al., 1999; Culter, 1999), for monitoring flavours and off-flavours in beer (Gardner et al., 1994; Pearce, 1994; Heberle et al., 2000) and for investigations into boar taint (Bourrounet et al., 1995; Annor-Frempong et al., 1998). Some workers have investigated the application of an electronic nose to the problem of contamination of corks by 2,4,6-trichloroanisole and have suggested that the system could provide a simple quality control tool (Rocha et al., 1998). There is some indication that electronic noses are beginning to be used in quality control. In the United Kingdom, a number of dairy companies are using electronic noses to screen incoming milk for taint, especially for chlorophenols and chlorocresols (Kilcast, 2002). Galdikas et al. (2000) used an array of eight MeOx sensors combined with a neural network to monitor chicken meat during storage. An electronic nose based on eight metalloporphyrines coated onto quartz microbalances (QMBs) was tested in the monitoring of ageing cod fillets and veal (Di Natale et al., 1997).

An alternative approach to sample introduction was described by Marsili (1999a; 2000), who used SPME to collect volatiles from the headspace above milk. The volatiles were rapidly desorbed and introduced to a mass spectrometer. The spectra from the samples were processed using multivariate statistics. This system could discriminate between various types of off-flavour in milk (Marsili, 1999a) and could be used to give accurate predictions of milk shelf-life (Marsili, 2000).

Other applications of electronic nose technology in this field include:

- classification of food spoilage bacteria (Pavlou et al., 2000; Magan et al., 2001)
- fish freshness (Di Natale et al., 1997)
- meat freshness (Eklov et al., 1998; Blixt and Borch, 1999)
- detecting off-odours in sugar (Kaipainen et al., 1997)
- taints in beer (Pearce et al., 1993; Tomlinson et al., 1995).

3.8 References

ACREE T E (1993), 'Gas chromatography–olfactometry', in Ho C T and Manley C H (eds), *Flavor Measurement*, New York, Marcel Dekker, pp 77–94.

AKASAKA K, OHTA H, HANADA Y and OHRUI H (1999), 'Simultaneous determination of lipid hydroperoxides by HPLC-post column systems', *Biosci, Biotechnol Biochem*, **63** (8), 1506–8.

ANKLAM E, BERG H, MATHIASSON L, SHARMAN M and ULBERTH F (1998), 'Supercritical fluid extraction (SFE) in food analysis: a review', *Food Additives Contamin*, **15** (6), 729–50.

ANNOR-FREMPONG I E, NUTE G R, WOOD J D, WHITTINGTON F W and WEST A (1998), 'The measurement of responses to different odour intensities of "boar taint" using a sensory panel and an electronic nose', *Meat Sci*, **50**, 139–51.

ARTHUR C L and PAULISZYN J (1990), 'Solid phase microextraction with thermal desorption using silica optical fibers', *Anal Chem*, **62**, 2145–8.

ARTHUR C L, KILLAM L M, MOTLAGH S, LIM M, POTTER D W and PAULISZYN J (1992), 'Analysis of substituted benzene compounds in groundwater using solid phase microextraction', *Environ Sci Technol*, **26**, 979–83.

AU-YEUNG C Y and MACLEOD A J (1980), 'Modification of the Likens and Nickerson apparatus to enable efficient extraction of aroma compounds from lipid samples', *Chem Ind*, **24**, 932–2.

AUNG L H, SMILANICK J L, VAIL P V, HARTSELL P L and GOMEZ E (1996), 'Investigations into the origin of chloroanisoles causing musty off-flavour in raisins', *J Agric Food Chem*, **44** (10), 3294–6.

BAYONOVE C and LEROY F (1994), 'Study on a method for the detection of chlorophenols in wines and corks', *Industrie delle Bevande*, **23** (131) 231–7.

BERRIE P G (1997), 'Fuzzy logic in the evaluation of sensor data', Kress-Rogers E (ed), in *Handbook of Biosensors and Electronic Noses: Medicine, Food and the Environment*, Boca Raton, CRC Press, pp 469–500.

BLANK I (1997), 'Gas chromatography–olfactometry in food aroma analysis', in Marsili R (ed), *Techniques for Analyzing Food Aroma*, New York, Marcel Dekker, pp 293–329.

BLIXT Y and BORCH E (1999), 'Using an electronic nose for determining the spoilage of vacuum-packaged beef', *Internat J Food Microbiol*, **46**, 123–34.

BORÉN H, GRIMVALL A, PALMBORG J, SÄVENHEAD R and WIGILIUS B (1985), 'Optimization of the open stripping system for the analysis of trace organics in water', *J Chromatogr*, **348**, 67–78.

BÖRJESSON T S, STOLLMAN W M and SCHNÜRER J L (1993), 'Off-odorous compounds produced by molds on oatmeal agar: identification and relation to other growth characteristics', *J Agric Food Chem*, **41** (11), 2104–11.
BOSSET J O, EBERHARD P, GALLMAN P, GAUCH R, RATTRAY W and SIEBER R (1996), 'Occurrence and behaviour of volatile sulfur-containing compounds in milk by heat treatment', in *Heat Treatments and Alternative Methods: Proceeding of a Symposium, Vienna, September 1995*, International Dairy Federation, Brussels, pp 409–21.
BOURROUNET B, TALOU T and GASET A (1995), 'Application of a multi-gas-sensor device in the meat industry for boar-taint detection', *Sensors Actuators B*, **27**, 250–4.
BRUNTON N P, CRONIN D A, MONAHAN F J and DURCAN R (2000), 'A comparison of solid-phase microextraction (SPME) fibres for measurement of hexanal and pentanal in cooked turkey', *Food Chem*, **68** (3), 339–45.
BUSER H-D, ZANIER C and TANNER H (1982), 'Identification of 2,4,6–trichloroanisole as a potent compound causing taint in wine', *J Agric Food Chem*, (**30**), 359-61.
BUTTERY R G, GUADAGNI D G, and LING L C (1976), 'Geosmin, a musty off-flavour of dry beans', *J Agric Food Chem*, **24** (2), 419–20.
BUTZKE C E, EVANS T J and EBELER S E (1998), 'Detection of cork taint in wine using automated solid-phase microextraction in combination with GC/MS–SIM', in Waterhouse A L (ed), *Chemistry of Wine Flavor: Developed from a Symposium, San Francisco, April 1997*, Washington DC, American Chemical Society, pp 208–16.
CAMEL V (1998), 'Supercritical fluid extraction of organic compounds', *Annal Falsifications de L'expertise Chim Toxologiq*, (**944**), 269–83.
CANTERGIANI E, BREVARD H, KREBS Y, FERIA-MORALES A, AMADO R and YERETZIAN C (2001), 'Characterization of the aroma of green Mexican coffee and identification of mouldy/earthy defect', *Eur Res Tech A*, 212, 648–57.
CAPONE S, SICILIANO P, QUARANTA F, RELLA R, EPIFANI M and VASANELLI L (2000), 'Analysis of vapours and foods by means of an electronic nose based on a sol-gel metal oxide sensors array', *Sensors Actuators B*, **69** (3), 230–5.
CHAINTREAU A (2001), 'Simultaneous distillation-extraction: from birth to maturity – review', *Flavour Fragrance J*, **16** (2), 136–48.
CHESTER T L, PINKSTON J D and RAYNIE D E (1998), Supercritical fluid chromatography and extraction' *Anal Chem*, **70** (12), 301R–319R.
CONTE E D, SHEN C-Y, PERSCHBACHER P W and MILLER D W (1996a), 'Determination of geosmin and methyl isoborneol in catfish tissue (*Ictalus punctatus*) by microwave-assisted distillation-solid phase adsorbent trapping', *J Agric Food Chem*, **44** (3) 829–35.
CONTE E D, SHEN C-Y, MILLER D W and PERSCHBACHER P W (1996b), 'Microwave distillation-solid phase adsorbent trapping device for the determination of off-flavours, geosmin and methyl isoborneol, in catfish tissue', *Anal Chem*, **68** (15), 2713–16.
CULTER J D (1999), 'The control of product and package quality with the electronic nose', *Tappi J*, 82, 194–200.
DALEY J D, LLOYD G T, RAMSHAW E H and STARK W (1986), 'Off-flavours related to the use of sorbic acid as a food preservative', *CSIRO Food Res Q*, **46**, 59–63.
DARRIET P, PONS M, LAMY S and DUBOURDIEU D (2000), 'Identification and quantification of geosmin, an earthy odorant contaminating wines', *J Agric Food Chem*, **48** (10), 4835–8.
DE HOFFMANN E and STROOBANT V (2001*)*, *Mass Spectrometry: Principles and Applications,* 2nd edn, Chichester, Wiley.
DI NATALE C, MACAGNANO A, DAVIDE F, D'AMICO A, PAOLESSE R, BOSCHI T, FACCIO M and FERRI G (1997), 'An electronic nose for food analysis', *Sensors Actuators B*, **44**, 521–6.

DITTMANN B and NITZ S (2000), 'Strategies for the development of reliable QA/QC methods when working with mass spectrometry-based chemosensory systems', *Sensors Actuators B*, **69** (3), 253–7.

DITTMANN B, HORNER G and NITZ S (1988), 'Development of a new chemical sensor on a mass spectrometric basis', *Adv Food Sci*, **20** (3–4), 115.

DOLLMANN B, WICHMANN D, SCHMITT A, KOEHLER H and SCHRIER P (1996), 'Quantitative analysis of 2-aminoacetophenone in off-flavoured wines by stable isotope dilution assay', *J Assoc Offic Anal Chem Internat*, **79** (2), 583–6.

DUPUY H P, FLICK JR G J, ST ANGELO A J and SUMRELL G (1986), 'Analysis for trace amounts of geosmin in water and fish', *J Am Oil Chem Soc*, **63** (7), 905–8.

DURST G L and LAPERLE E A (1990), 'Styrene monomer migration as monitored by purge and trap gas chromatography and sensory analysis for polystyrene containers', *J Food Sci*, **55** (2), 522–4.

EKLOV T, JOHANSSON G, WINQUIST F and LUNDSTROM I (1998), 'Monitoring sausage fermentation using an electronic nose', *J Sci Food Agric*, **76**, 525–32.

ELMORE J S, ERBAHADIR M A and MOTTRAM D S (1997), 'Comparison of dynamic headspace concentration on Tenax with solid phase microextraction for the analysis of aroma volatiles', *J Agric Food Chem*, **45** (7), 2638–41.

ESCUDERO A, HERNANDEZ-ORTE P, CACHO J and FERREIRA V (2000), 'Clues about the role of methional as character impact odorant of some oxidized wines', *J Agric Food Chem*, **48** (9), 4268–72.

EVANS P, PERSAUD K C, MCNEISH A S, SNEATH R W, HOBSON N and MAGAN N (2000), 'Evaluation of a radial basis function neural network for the determination of wheat quality from electronic nose data', *Sensors Actuators B*, **69** (3), 348–58.

EVANS T J, BUTZKE C E and EBELER S E (1997), 'Analysis of 2,4,6-trichloroanisole in wines using solid-phase microextraction coupled to gas chromatography–mass spectrometry', *J Chromatogr A*, **786** (2), 293–8.

FENG Y-W and ACREE T E (1999), 'Gas chromatography olfactometry in aroma analysis', *Foods Food Ingredients J Japan*, **179**, 57–66.

FISCHER C and FISCHER U (1997), 'Analysis of cork taint in wine and cork material at olfactory subthreshold levels by solid phase microextraction', *J Agric Food Chem*, **45** (6), 1995–7.

FORSGEN G, FRISELL H and ERICSSON B (1999), 'Taint and odour related to quality monitoring of two food packaging board products using gas chromatography, gas sensors and sensory analysis', *Nordic Pulp Paper Res J*, **14**, 5–16.

FRIEDRICH J E and ACREE T E (2000), 'Issues in gas chromatography–olfactometry methodologies', in *Flavour Chemistry: Industrial and Academic Research*, Washington D C, American Chemical Society, 124–32.

GALDIKAS A, MIRONAS A, SENULIENE D, STRAZDIENE V, SETKUS A and ZELININ D (2000), 'Response time based output of metal oxide gas sensors applied to the evaluation of meat freshness with neural signal analysis', *Sensors Actuators B*, **69** (3), 258–65.

GARCIA-REGUEIRO J A and RUIS M A (1998), 'Rapid determination of skatole and indole in pig back fat by normal phase liquid chromatography', *J Chromatogr A*, 809 (1–2), 246–51.

GARDNER J W and BARTLETT P N (1999), *Electronic Noses: Principles and Application*, Oxford University Press.

GARDNER J W and HINES E L (1997), 'Pattern analysis techniques', in Kress-Rogers E (ed), *Handbook of Biosensors and Electronic Noses: Medicine, Food and the Environment*, Boca Raton, CRC Press, pp 633–52.

GARDNER J W, PEARCE T C, FRIEL S, BARTLETT P N and BLAIR N (1994), 'A multisensor system for beer flavour monitoring using an array of conducting polymers and predictive classifiers', *Sensors and Actuators B*, **18–19**, 240–3.

GINZBURG B, CHALIFA I, ZOHARY T, HADAS O, DOR I and LEV O (1998), 'Identification of oligosulfide odorous compounds and their sources in the Lake of Galilee', *Water Res*, **32** (6), 1789–800.
GOLASZEWSKI R, SIMS C A, O'KEEFE S F, BRADDOCK R J and LITTELL R C (1998), 'Sensory attributes and volatile components of stored strawberry juice', *J Food Sci*, **63** (4), 734–8.
GRIMM C C, LLOYD S W, BATISTA R and ZIMBA P V (2000a), 'Using microwave distillation solid-phase microextraction gas chromatography–mass spectrometry for analysing fish tissue', *J Chromatogr Sci*, **38** (7), 289–96.
GRIMM C C, LLOYD S W, ZIMBA P V and PALMER M (2000b), 'Microwave distillation solid-phase microextraction gas chromatographic analysis of geosmin and 2-methylisoborneol in catfish', *Am Lab*, **2**, 40–8.
GROB K (1973), 'Organic substances in potable water and in its precursor: Part I Methods for their determination by gas-liquid chromatography', *J Chromatogr*, **84**, 255–73.
GROB K and ZÜCHER F (1976), 'Stripping of trace organic substances from water – equipment and procedure', *J Chromatogr*, **117**, 285–94.
HANAOKA K, VALLET N, GIAMPAOLI P and MACLEOD P (2001), 'Possible influence of breathing on detection frequency and intensity rating in gas chromatography–olfactometry', *Food Chem*, **72** (1), 97–103.
HANAOKO K, SIEFFERMANN J-M and GIAMPAOLI P (2000), 'Effects of the sniffing port air makeup in gas chromatography–olfactometry', *J Agric Food Chem*, **48** (6), 2368–71.
HANDLEY A J and ALARD E R (2001), *Gas Chromatographic Techniques and Applications*, Sheffield, Sheffield Academic Press.
HARRIS N D, KARAHADIAN C and LINDSAY R C (1986), 'Musty aroma compounds produced by selected molds and actinomycetes on agar and whole wheat bread', *J Food Protein*, **49** (12), 964–70.
HEBERLE I, LIEBMINGER A, WEIMAR U and GÖPEL W (2000), 'Optimized sensor arrays with chromatographic preseparation: characterisation of alcoholic beverages', *Sensors Actuators B*, **68**, 53–7.
HEIKES D L and GRIFFITT K R (1980), 'Gas-liquid chromatographic determination of pentachlorophenol in Mason jar lids and home canned foods', *J Assoc Off Anal Chem*, **63** (5), 1125–7.
HILL P G and SMITH R M (2000), 'Determination of sulfur compounds in beer using headspace solid-phase microextraction and gas chromatographic analysis with pulsed flame photometric detection', *J Chromatogr A*, **872** (2), 203–13.
HOCKING A D, SHAW K J, CHARLEY N J and WHITFIELD F B (1998), 'Identification of an off-flavour produced by *Penicillium solitum* in margarine', *J Food Mycology*, **1** (1), 23–30.
HOEFER U, BÖTTNER H, FELSKE A, KÜHNER G, STEINER K and SULZ G (1997), 'Thin-film SnO_2 sensor arrays controlled by variation of contact potential – a suitable tool for chemometric gas mixture analysis in the TLV range', *Sensors Actuators B*, **44**, 429–33.
HOLSCHER W, BADE-WEGNR H, BENDIG I, WOLKENHAUER P and VITZTHUM O G (1995), 'Off-flavour elucidation in certain batches of Kenyan coffee', 16th International Scientific Colloquium on Coffee, Paris, Association Scientifique International de Café pp 174–82.
HORWOOD J F, LLOYD G T, RAMSHAW E H and STARK W (1981), 'An off-flavour associated with the use of sorbic acid during feta cheese maturation', *Austral J Dairy Techol*, **36** (1), 38–40.
HSIEH T C-Y, TANCHOTIKUL U and MATIELLA J E (1988), 'Identification of geosmin as the major muddy off-flavour of Louisiana brackish water clam (*Raugia cuneata*)', *J Food Sci*, **53** (4), 1228–9.

HUBSCHMANN H-J (2001), *Handbook of GC/MS: Fundamentals and Applications*, Weinheim, Wiley.

HUCK P M, KENEFICK S L, HRUDEY S E and ZHANG S (1995), 'Bench-scale determination of the removal of odour compounds with biological treatment', *Water Sci Technol*, **31** (11), 203–9.

HUO L H, LI X L, LI W and XI S Q (2000), 'Gas sensitivity of composite Langmuir–Blodgett films of Fe_2O_3 nanoparticle-copper phthalocyanine', *Sensors and Actuators B*, **68**, 77–81.

HUOTARI M J (2000), 'Biosensing by insect olfactory receptor neurons', *Sensors Actuators B*, **71**, 212–22.

KAIPAINEN A, YLISUUTARI S, LUCAS Q and MOY L (1997), 'A new approach to odour detection', *Internat Sugar J*, **99**, 403–8.

KILCAST D (2002), Personal communication.

KLEYKERS R W G and SCIFFERSTEIN H N J (1995), 'Methodological issues in GC–olfactometry', *Voedingsmiddelentechnologie*, **28** (21), 26–9.

KOGA Y, KOGA T, SAKAMOTO K and OHTA H (2001), 'Influence of off-flavour formed from plasma protein on flavour characteristics of emulsion-type dressing', *J Jpn Soc Food Sci Technol*, **48** (7), 507–13.

KOHL D (1997), 'Semiconductor and Calorimetric Devices and Arrays', in Kress-Rogers E (ed), *Handbook of Biosensors and Electronic Noses: Medicine, Food and the Environment*, Boca Raton, CRC Press, pp 533–61.

KOTSERIDIS Y, RAY J-L, AUGIER C and BAUMES R (2000), 'Quantitative determination of sulfur containing wine odorants at sub-ppb levels. 1. Synthesis of the deuterated analogues', *J Agric Food Chem*, **48** (12), 5819–23.

KRASNER S W, HWANG S J and MCGUIRE M J (1983), 'A standard method for quantification of earthy-musty odorants in water, sediments and algal cultures', *Water Sci Technol*, **15**, 127–38.

LAMPE U, FLEISCHER M, REITMEIER N, MEIXNER H, MCMONAGLE J B and MARSH A (1996), 'New metal oxide sensors: materials and properties', in Baltes H, Göpel W and Hesse J (eds), *Sensors Update Volume 2*, Weinheim, Wiley-VCH, pp 1–36.

LAVIGNE-DELCROIX A, TUSSEAU D and PROIX M (1996), 'Validation of a chromatographic chemiluminescence detector', *Sci Aliments*, **16** (3), 267–80.

LAYE I, KARLESKIND D and MORR C V (1995), 'Dynamic headspace analysis of accelerated storage commercial whey protein concentrate using four different adsorbent traps', *Milchwissenschaft*, **50** (5), 268–72.

LE BIZEC B, MONTRADE M-P and ANDRE F (2000), *Gas Chromatography*, Amsterdam, Harwood Academic.

LEE H S and NAGY S (1990), 'Formation of 4-vinyl guaiacol in adversely stored orange juice as measured by an improved HPLC method', *J Food Sci*, **55** (1), 162–3, 166.

LEE M-R, LEE R-J, LIN Y-W, CHEN C-M and HWANG B-H (1998), 'Gas-phase postderivitisation following solid-phase microextraction for determining acidic herbicides in water', *Anal Chem*, **70** (9), 1963–8.

LIKENS S T and NICKERSON G B (1964), 'Detection of certain hop oil constituents in brewing products', *Am Chem Soc Brew Proc*, 5–13.

LLOYD S W and GRIMM C C (1999), 'Rapid analysis of geosmin and 2-methylisoborneol in catfish using microwave desorption / solid phase-micro-extraction procedures'. *J Agric Food Chem*, **47** (1), 164–9.

LLOYD S W, LEA J, ZIMBA P and GRIMM C C (1998), 'Rapid analysis of geosmin and 2-methylisoborneol in water using SPME procedures', *Water Res* **32** (7), 2140–6.

MAGAN N, PAVLOU A and CHRYSANTHAKIS I (2001), 'Milk-sense: a volatile sensing system recognises spoilage bacteria and yeasts in milk', *Sensors Actuators B*, **72** (1), 23–34.

MAGARD M A, BERG H E B, TAGESSON V, JAREMO M L G, KARLSSON L L H, MATHIASSON L J E, BONNEAU M and HANSE-MOELLER J (1995), 'Determination of androstenone in pig fat using supercritical fluid extraction and gas chromatograph–mass spectrometry', *J Agric Food Chem*, **43** (1), 114–20.

MARSILI R T (1999), 'Comparison of solid-phase microextraction and dynamic headspace methods for the gas chromatographic–mass spectrometric analysis of light-induced lipid oxidation products in milk', *J Chromatogr Sci*, **37** (1), 17–23.

MARSILI R T (1999a), 'SPME–MS–MVA as an electronic nose for the study of off-flavors in milk', *J Agric Food Chem*, **47** (2), 648–54.

MARSILI R T (2000), 'Shelf-life prediction of processed milk by solid-phase microextraction, mass spectrometry and multivariate analysis', *J Agric Food Chem*, **48** (8), 3470–5.

MARSILI R T, MILLER N, KILMER G J and SIMMONS R E (1994), 'Identification and quantitation of the primary compounds responsible for the characteristic malodor of beet sugar by purge and trap GC-MS-OD techniques', *J Chromatogr Sci*, **32** (1), 165–71.

MARTH E H, CONSTANCE M, CAPP C M, HASENZAHL L, JACKSON H W and HUSSONG H V (1966), 'Degradation of potassium sorbate by *Penicillium* species', *J Dairy Sci*, **49**, 1197–9.

MARTIN J F, FISHER T H and BENNET L W (1988), 'Musty odor in chronically off-flavored channel catfish: isolation of 2-methylenebornane and 2-methyl-2-borneol', *J Agric Food Chem*, **36** (6), 1257–60.

MAZZA G and PIETZAK E M (1990), 'Headspace volatiles and sensory characteristics of earthy, musty flavoured potatoes', *Food Chem*, **36** (2), 97–112.

MCCALLUM R, PENDLETON P, SCUMANN R and TRINH M (1998), 'Determination of geosmin and 2-methylisoborneol in water using SPME and GC-chemical ionization/electron impact ionization ion trap MS', *Analyst*, **123** (10), 2155–60.

MCGORIN R J, POFAHL T R and CROASMUN W R (1987), 'Identification of the musty component from an off-odor packaging film', *Anal Chem*, **59** (18), 1109A–1112A.

MESTRES M, BUSTO O and GUASCH J (1997), 'Chromatographic analysis of volatile sulfur compounds in wines using the static headspace technique with flame photometric detection', *J Chromatogr A*, **773** (1–2), 261–9.

MESTRES M, BUSTO O and GUASCH J (1998), 'Headspace solid-phase microextraction analysis of volatile sulfides and disulfides in wine aroma', *J Chromatogr A*, **808** (1–2), 211–18.

MESTRES M, MARTI M P, BUSTO O and GUASCH J (1999a), 'Simultaneous analysis of thiols, sulfides and disulfides in wine aroma by headspace solid-phase microextraction–gas chromatography', *J Chromatogr A*, **849** (1), 293–7.

MESTRES M, SALA C, MARTI M P, BUSTO O and GUASCH J (1999b), 'Headspace solid-phase microextraction of sulfides and disulfides using Carboxen-polydimethylsiloxane fibers in the analysis of wine aroma', *J Chromatogr A*, **835** (1), 137–44.

MESTRES M, BUSTO O and GUASCH J (2000), 'Analysis of organic sulfur compounds in wine aroma', *J Chromatogr A*, **881** (1–2), 569–81.

MIELLE P and MARQUIS F (2000), 'Gas sensors arrays ('Electronic Noses'): a study about the speed/accuracy ratio', *Sensors Actuators B*, **68**, 9–16.

MIELLE P, MARQUIS F and LATRASSE C (2000), 'Electronic noses: specify or disappear', *Sensors Actuators B*, **68**, 287–94.

MILLS O E, GREGORY S P, VISSER F R and BROOME A J (1997), 'Chemical taint in rindless Gouda cheese', *J Agric Food Chem*, **45** (2), 487–92.

MISTRY B S, REINECCIUS G A and JASPER B L (1994), 'Comparison of gas chromatographic detectors for the analysis of volatile sulfur compounds in food', in Muissinan C J, and Keelan M E (eds), *Sulfur Compounds in Foods: Proceedings of a Symposium, Chicago, August 1993*, Washington D C, American Chemical Society, pp 8–21.

MOIO L, DIANA M, DEL PRETE G, DI LANDE G, VALENTINO A A and BIANCO A (1998), 'GC-sniffing and GC-MS study of volatiles of "normal" corks, corks affected by "yellow stain"; from *Quercus Suber L.* and wines with defect of cork odour', *Industrie delle Bevande*, **27** (158), 615–19.

MONTI S M, RITIENI A, GRAZIANI G, RANDAZZO G, MANNINA L, SEGRE A L and FOGLIANO V (1999), 'LC/MS analysis and antioxidative efficiency of Maillard reaction products from a lactose–lysine model system', *J Agric Food Chem*, **47** (4) 1506–13.

MORALES M T, RIOS J J and APARICIO R (1997), 'Changes in the volatile composition of virgin olive oil during oxidation: flavors and off-flavors', *J Agric Food Chem*, **45** (7), 2666–73.

MUKHOPADHYAY M (2000), 'Fundamentals of supercritical fluids and phase equilibria' in Mukhopadhyay M (ed), *Natural Extracts Using Supercritical Carbon Dioxide*, Boca Raton, CRC Press, 11–82.

NEDJMA M and MAUJEAN A (1995), 'Improved chromatographic analysis of volatile sulfur compounds by the static headspace technique on water–alcohol solutions and brandies with chemiluminescence detection', *J Chromatogr*, **704** (2), 495–502.

OCHIAI N, SASAMOTO K, TAKINO M, YAMASHITA S, DAISHIMA S, HEIDEN A and HOFFMAN A (2001), 'Determination of trace amounts of off-flavour compounds in drinking water by stir bar sorptive extraction and thermal desorption GC-MS', *Analyst*, **126** (10), 1652–7.

OFFEN C, SQUIBB A and PATEL P (2001), 'New extraction and concentration techniques for detecting low levels of chemical tainting compounds in foodstuffs', *Leatherhead Food Research Association Technical Reports*, 777.

PAASIVIRTA J and KOISTINEN J (1994), 'Chlorinated ethers', in Kicenuik J W and Ray S (eds), *Analysis of Contaminants in Edible Aquatic Resources: General Considerations, Metals, Organometallics, Tainting and Organics*, Weinheim, VCH, 411–27.

PATTERSON R L S (1970), 'Detection of Meat Odours', *Process Biochem*, **5** (5), 27–31.

PAULISZYN J (2000), 'Theory of solid-phase microextraction', *J Chromatogr Sci*, **38** (7), 270-8.

PAVLOU A, MAGAN N, SHARP D, BROWN J, BARR H and TURNER A P F (2000), 'An *in vitro* rapid odour detection and recognition model in discrimination of *H. pylori* and other gastroeosophageal pathogens', *Biosensors Bioelectron*, **15** (7–8), 333–42.

PEARCE T C, GARDNER J W, FRIEL S, BARTLETT P N and BLAIR N (1993), 'Electronic nose for monitoring the flavour of beers', *Analyst*, **118**, 371–7.

PERSAUD K C and DODD G H (1982), 'Analysis of discrimination mechanisms in the mammalian olfactory system using a model nose', *Nature*, **299**, 352–5.

PERSAUD K C and TRAVERS P J (1997), 'Arrays of broad specificity films for sensing volatile chemicals', in Kress-Rogers E (ed), *Handbook of Biosensors and Electronic Noses: Medicine, Food and the Environment*, Boca Raton, CRC Press, pp 563–92.

PETERSEN M A, POLL L and LARSEN L M (1999), 'Identification of compounds contributing to boiled potato off-flavour', *Lebens-Wiss u-Technol*, **32** (1), 32–40.

PITT P (1996), 'A nose for a problem (odour and taint detection systems)', *Packaging Week*, **12** (18), 33.

POTTER D W and PAULISZYN J (1992), 'Detection of substituted benzene in water at the pg/ml level using solid phase microextraction and gas chromatography–ion trap mass spectrometry', *J Chromatogr*, **625**, 247–55.

RAMSTAD T and WALKER J S (1992), 'Investigation of musty odour in pharmaceutical products by dynamic headspace gas chromatography', *Analyst*, **117**, 1361–6.

RAPP A (1998), 'Volatile flavour of wine: correlation between instrumental analysis and sensory perception', *Nahrung*, **42** (6), 351–63.

REN Y L, DESMARCHELIER J M, WILLIAMS P and DELVES R (2001), 'Natural levels of dimethyl sulfide in rough rice and its products', *J Agric Food Chem*, **49** (2), 705–9.

ROCHA S, DELGADILLO I, CORREIA A J F, BARROS A and WELLS P (1998), 'Application of an electronic aroma sensing system to cork stopper quality control', *J Agric Food Chem*, **46** (1), 145–51.

RUIS M A and GARCIA-REGUEIRO J A (2001), 'Skatole and indole concentrations in *Longissimus dorsi* and fat samples of pigs', *Meat Sci*, **59** (3), 285–91.
SANCHEZ SAEZ J J, GARRALETA M D, CALVO ANTON P and FOLGUEIRAS ALONSO M L (1991), 'Identification of 4-bromo-2-chloro-phenol as a contaminant responsible for organoleptic taint in melons', *Food Additives Contamin*, **8** (5), 627–32.
SANDERS R A and SEVENANTS M R (1994),'Identification of off-flavor components in juice beverages', poster presented at the *Proceedings of the 42nd American Society of Mass Spectrometry Conference*, Chicago, 1994.
SCARLATA C J and EBELER S E (1999), 'Headspace solid-phase microextraction for the analysis of dimethyl sulfide in beer', *J Agric Food Chem*, **47** (7), 2505–8.
SCHULTZ T H, FLATH R A, MON T R, EGGLING S B and TERANISHI R (1977), 'Isolation of volatile components from a model system', *J Agric Food Chem*, **25** (2), 446–9.
SCHÜTZ S, WEIBBECKER B and HUMMEL H E (1996), in *Refereed Abstracts of the Fourth World Congress on Biosensors, Bangkok, Thailand, May 1996*, Oxford, Elsevier, p 85.
SCHÜTZ S, WEIBBECKER B, HUMMEL H E, SCHÖNING M J, RIEMER A, KORDOS P and LÜTH H (1997), 'A field effect transistar-insect antenna junction', *Naturwissenschaften*, **84**, 86–8.
SENSIDONI A, RONDININI G, PERESSINI S, MAIFRENI M and BORTOLOMEAZZI R (1994), 'Presence of an off-flavour associated with the use of sorbates in cheese and margarine', *Ital J Food Sci*, **6** (2), 237–42.
SEVENANTS M R and SANDERS R A (1984), 'Anatomy of an off-flavour investigation: The "medicinal" cake mix', *Anal Chem*, **56** (2), 293A–298A.
SHIERS V P and SQUIBB A D (1999), 'Evaluation of the Hewlett-Packard chemical sensor HP4440A – Detection of taints in packaging', *Leatherhead Food Research Association Technical Notes*, 133.
SHIERS V, ARDECHY M and SQUIBB A (1999), 'A new mass-spectrometry based electronic nose for headspace characterization', in *Electronic Noses and Sensor Array Based Systems*, Technomic, Lancaster, pp 289–95.
SHIREY R E (2000), 'Optimization of extraction conditions and fiber selection for semivolatile analytes using solid-phase microextraction' *J Chromatogr Sci*, **38** (7), 279–88.
SIEGMUND B, LEITNER E, MAYER I, PFANNHAUSER W, FARKAS P, SADECKA J and KOVAC M (1997), '5,6-Dihydro-2,4,6-trimethyl-4*H*-1,3,5-dithiazine – an aroma-active compound formed in the course of the Likens Nickerson extraction', *Zeit Lebens– Unters–Forsch A*, **205** (1), 73–5.
SIHVONEN M, JARVENPAA E, HIETANIEMI V and HUOPALAHTI R (1999), 'Advances in supercritical carbon dioxide technologies', *Trends Food Sci Tech*, **10** (6–7), 217–22.
SIMPSON R F, AMON J M and DAW A J (1986), 'Off-flavour in wine caused by guaiacol' *Food Technol Austral*, **38** (1), 31–3.
SIOUFFI A M and DELAURENT C (1996), 'Instruments and techniques', in Nollet L M L (ed), *Handbook of Food Analysis, Volume 2: Residues and Other Food Component Analysis*, New York, Marcel Dekker, pp 1907–35.
SPADONE J-C, TAKEOLA G and LIARDON R (1990), 'Analytical investigations of Rio off-flavour in green coffee', *J Agric Food Chem* **38**, 226–33.
STEELY J S (1994), 'Chemiluminescence detection of sulfur compounds in cooked milk', in Mussinan C L and Keelan M E (eds), *Sulfur Compounds in Foods: Proceedings of a Symposium, Chicago, August 1993*, Washington D C, ACS, pp 22–5.
STEPHAN A and STEINHART H (1999), 'Identification of character impact odorants of different soybean lecithins', *J Agric Food Chem*, **47** (7), 2854–9.
SWAN H B (2000), 'Determination of existing and potential dimethyl sulfide in red wines by gas chromatography atomic emission spectroscopy', *J Food Compos Anal*, **13** (3), 207–17.

TAYLOR M K, YOUNG T M, BUTZKE C E and EBELER E (2000), 'Supercritical fluid extraction of 2,4,6-trichloroanisole from cork stoppers', *J Agric Food Chem*, **48** (6) 2208–11.
THEISEN M, STEUDEL A, RYCHETSKY M and GLESNER M (1998), 'Fuzzy logic and neurosystems assisted intelligent sensors', in Baltes H, Göpel W and Hesse J (eds), *Sensors Update Volume 3*, Weinheim, Wiley-VCH, pp 29–59.
TOMLINSON J, ORMROD H and SHARPE F (1995), 'Electronic aroma detection in the brewery', *J Am Soci Brewing Chem*, **53**, 167–73.
TURNES M I, RODRGUEZ I, MEJUTO M C and CELA R (1994), 'Determination of chlorophenols in drinking water samples at the subnanogram per millilitre level by gas chromatography with atomic emission detection', *J Chromatogr*, **683** (1), 21–9.
VAN AARDT M, DUNCAN S E, BOURNE D, MARCY J E, LONG T E, HACKNEY C R and HEISEY C (2001), 'Flavor threshold for acetaldehyde in milk, chocolate milk and spring water using solid phase microextraction gas chromatography for quantification', *J Agric Food Chem*, **49** (3), 1377–81.
VAN RUTH S M and O'CONNOR C H (2001), 'Influence of assessors' qualities and analytical conditions on gas chromatography–olfactometry analysis', *Eur Food Res Technol*, **213** (1), 77–82.
WELLNITZ-RUEN W, REINECCIUS G A and THOMAS E L (1982), 'Analysis of the fruity off-flavour in milk using headspace concentration capillary column gas chromatography', *J Agric Food Chem*, **30** (3), 512–14.
WHITFIELD F B, FREEMAN D J and SHAW K J (1983), ' Trimethylarsine; an important off-flavour component in some prawn species', *Chem Ind*, **20**, 786–7.
WHITFIELD F B, LAST J H, SHAW K J and TINDALE C R (1988), '2,6-Dibromophenol: the cause of an iodoform-like off-flavour in some Australian crustacea', *J Sci Food Agric*, **46** (1), 29–42.
WHITFIELD F B, SHAW K J and LY NGUYEN T H (1986), 'Simultaneous determination of 2,4,6-trichloroanisole, 2,3,4,6-tetrachloroanisole and pentachloroanisole in foods and packaging materials by high resolution gas chromatography–multiple ion monitoring mass spectrometry', *J Agric Food Chem* **37**, 85–96.
WHITFIELD F B, SHAW K J, GIBSON A M and MUGFORD D C (1991), 'An earthy off-flavour in wheat flour: geosmin produced by *Streptomyces griseus*', *Chem Ind*, **22**, 841–2.
WHITFIELD F B, HILL J L and SHAW K J (1997), '2,4,6-Tribromoanisole: a potential cause of mustiness in packaged food', *J Agric Food Chem*, **45** (3), 889–93.
WHITFIELD F B, JENSEN N and SHAW K J (2000), 'Role of *Yersinia intermedia* and *Pseudomonas putida* in the development of a fruity off-flavour in pasteurized milk', *J Dairy Res*, **67** (4), 561–9.
WILKES J G, CONTE E D, KIM Y, HOLCOMB M, SUTHERLAND J B and MILLER D W (2000), 'Sample preparation for the analysis of flavours and off-flavours in foods', *J Chromatogr A*, **880**, 3–33.
YAMPRAYOON J and NOOHORM A (2000a), 'Geosmin and off-flavour in Nile tilapia (*Oreochromis niloticus*)', *J Aquatic Food Prod Technol* **9** (2), 29–41.
YAMPRAYOON J and NOOHORM A (2000b), 'Effects of preservation methods on geosmin content and off-flavour in Nile tilapia (*Oreochromis niloticus*)', *J Aquatic Food Prod Technol*, **9** (4), 95-107.
YUOMOLA M, VAHVA M and KALLIO H (1996), 'High-performance liquid chromatographic determination of skatole and indole levels in pig serum, subcutaneous fat and submaxillary salivary glands', *J Agric Food Chem*, **44** (5), 1265–70.
ZABLOTSKY D A, PATERSON J A, FORREST J C, CHEN L F and GRANT A L (1993), 'Supercritical fluid extraction for boar taint measurement', *Meat Focus Internat*, **2** (11), 494–5.
ZABLOTSKY D A, CHEN L F, PATTERSON J A, FORREST J C, LIN H M and GRANT A L (1995), 'Supercritical carbon dioxide extraction of androstenone and skatole from pork fat', *J Food Sci*, **60** (5), 1006–8.

ZHOU A, MCFEETERS R F and FLEMING H P (2000), 'Development of oxidised odor and volatile aldehydes in fermented cucumber tissue exposed to oxygen', *J Agric Food Chem*, **48** (2), 193–7.

ZHU M, AVILES F J, CONTE E D, MILLER D W and PERSCHBACHER P W (1999), 'Microwave mediated distillation with solid-phase microextraction: determination of off-flavours, geosmin and methyl isoborneol, in catfish tissue', *J Chromatogr A*, **833**, 223–30.

4
Packaging materials as a source of taints

T. Lord, Pira International, UK

4.1 Introduction

The packaging used with a food has become a critical part of the food product. Apart from containing and protecting the food, it also provides a brand image enabling the product to catch the eye and stand out from its competitors. In the case of convenience foods it also provides functionality, for example it may be a tray that the food is eaten from or a baking sheet that browns the food in the microwave oven. The primary purpose of the packaging is, of course, to deliver the food at its highest quality to the consumer. Unfortunately, although packaging is used to protect the food from contamination and spoilage, it can also occasionally be a source of tainting substances.

Taint problems, although rare in comparison to the huge volume of packaging produced, are sometimes serious with significant financial losses involved. A slightly unpleasant odour or taste may be sufficient for a consumer to reject the food. Even an unpleasant odour from the packaging upon opening the packet may be sufficient for the consumer to reject the product. The concentrations of chemicals required to cause rejection may be very low, in some cases in the parts per trillion (ppt). Generation of a taint may depend on a combination of factors, all of which have to be present simultaneously for the taint to occur. Whether a chemical will cause an unpleasant taste or odour depends on the food. If the tainting chemical gives a taste or odour in keeping with the food, a complaint may not occur. As an example of this, butyl acetate, which is a common ink solvent, has a fruit odour. This may be masked in a fruit drink. A fruit odour in a chocolate drink, however, may cause an unpleasant clash of aromas and not be acceptable. The ability of people to notice the taint varies widely. Occasionally a person who is insensitive to one particular taint may be especially sensitive to another. The ability to recognise a taint increases dramatically once the person has become familiar with the taint. To some people a particular taint may not be unpleasant, whilst to others it may be unacceptable. It is these factors that make the investigation and control of taint from packaging so difficult. For the analyst, the very low concentrations of chemicals required to cause a taint impose special difficulties. The purpose of this chapter is to provide an understanding of the common causes of taints from packaging.

4.1.1 Basic definitions

A taint is a taste or an odour or an undesirable change in the flavour resulting from the presence of one or more chemicals in the food. The standard definition of taint (ISO, 1992) is 'a taste or odour foreign to the product'. The definition also distinguishes an off-flavour as an atypical flavour usually associated with deterioration. An off-flavour is also defined as 'unpleasant odour or flavour imparted to food through internal deteriorating change'. Malodour is used as a term for the unwanted odour present in the food when no assumption of food deterioration has been made. It is helpful to use the following terminology:

- **Off-odour:** an atypical odour that results from a deterioration of the food
- **Malodour:** unwanted odour usually from contamination by a chemical foreign to the food
- **Taint:** an atypical taste or odour imparted to the food from the presence in the food of one or more chemicals.

An off-flavour or odour can also arise when compounds are lost from the food resulting in a change or imbalance in flavour and aroma.

The word 'taint' is a general term for taste and odour but it leads some people to think incorrectly in terms of a single human sense. It is incorrect to compare a taste threshold with an odour threshold and vice versa. After all we do not compare sight with sound. It is sometimes the case with packaging that only taste or odour are involved. It is best not to use the word taint when it is known that only an odour is involved. Some people think of taint as being associated only with taste giving rise to the term 'taint and odour'. This is not strictly true, as taint can be an odour or a taste. It is very useful in an investigation of taint to decide whether it is an odour or a taste or both. If it is only taste it usually means that the concentration of the chemical(s) responsible is at least one order of magnitude lower than if an odour is also involved. As a general rule, it is easier to investigate the cause of an odour than it is a taste. Odours usually involve chemicals that are volatile enough to partition into the air. Volatile compounds are much easier to isolate from the food matrix for chemical analysis, and are usually amenable to analysis by gas chromatography–mass spectrometry (GC–MS). When a complaint is made it is important for the person dealing with the complaint to determine whether there is an odour or a taste problem or both. An experienced analytical chemist will immediately then have an idea of the sort of techniques that must be used and the likely costs involved.

4.1.2 Commercial considerations

The cost of a taint problem can be significant. Invariably it means the disposal of all the packaging and the product. The chain in the packaging production process is usually complicated and extended. A packaging converter may only be one of numerous converters involved in production. The taint can be introduced at numerous different stages of production so that identifying the entry point can be very difficult.

Retailers and converters have a better chance of solving a taint problem if they involve an independent consultant with extensive experience in the field. There are numerous laboratories experienced in the investigation of taints. The majority of these operate in competition with each other ensuring that a good value professional service is readily available. It is seldom cost effective for a supplier to maintain the technical resources required to investigate taint problems. It is the experience of the author that the cause of a packaging related taint is often only identified by a very experienced analytical chemist using state-of-the-art analytical equipment, working with the converter, packer or retailer who also has the necessary commitment to see the problem solved.

4.2 Main types of food packaging

Packaging can be grouped into a number of basic types. These types are discussed below:

- plastics and plastic laminates
- glass
- paper board
- paper plastic laminates and metal composites
- metal
- regenerated cellulose films.

4.2.1 Plastics and plastic laminates

Plastic laminates here include all polymer laminates as well as paper-to-metal laminates. Laminates are used because, in spite of recent advances in polymer science, there is still no one polymer type that meets all the cost and performance requirements needed for food packaging. The layers in a laminate are bonded together using adhesives, which can contain solvents. In addition, the laminates may be printed with solvent-based inks. Additives are added to the plastics to protect them from thermal degradation during processing into sheets, bottles etc. All of these chemicals, if present in the final packaging material, can sometimes migrate into the packaged food, and then may occasionally give rise to a taint.

Common polymers include polyethylene (PE), linear lowdensity (LLDPE), lowdensity (LDPE), highdensity (HDPE), polypropylene (PP), polystyrene (PS), polyvinyl chloride (PVC) and polyethylene terephthalate (PET). The ease of migration of many tainting compounds through these polymers follows the order of polymers given, with the greatest migration through PE and the least through PET. In general, the more crystalline the polymer, the better the barrier properties. Typically, polyethylene is the polymer that is in contact with the food, with higher melting point polymers, barrier layers and coatings on the outer surface. Often the print has an outer layer applied to it. This protects the print and gives a gloss image. The result is that any tainting component is likely to migrate backwards into the food rather than out through the higher barrier polymers on the outer surface. Polyolefins (polyethylene or polypropylene) can also affect the taste of packaged food and drink by removing flavour components. This is called aroma scalping. Marin *et al.* (1992) analysed packaged orange juice. They showed the absorption of key components of orange juice into polyolefins.

4.2.2 Glass
Glass is one of the packaging types least likely to give taint problems. Taints from glass packaging usually occur as a result of contamination of the food from a component or contaminant in the closures, or a contaminant deposited on the inside of the glass container. Contamination from the closure may occur from the metal lacquers used or the polymeric seal. Glass is often coated with thin coatings of polyethylene or aliphatic esters. This protects the glass from scuffing and also confers a lubricating or slip property to the surface, for example on screw threads. These coatings may, on rare occasions, degrade and give rancid odours.

4.2.3 Paperboard
Paper is used as secondary packaging, for example for breakfast cereal, where a plastic inner bag is used to maintain the shelf-life of the product and the paperboard box provides the aesthetic and brand appearance as well as pack handling convenience. Migration of components from the paperboard into food in such packaging is usually low. Where taints do occur, they usually result from the contamination of the paper carton from

printing inks and solvents, or a chemical in the coatings or adhesives. Sizing agents (agents that control the absorbtivity of the paper) and wet strength agents are added to paper. These very rarely cause taint. The majority of paper used for packaging has some form of coating applied. The coating serves to provide a good surface for printing and thereby improves the appearance of the packaging. The most common coatings on paper consist of filler, a pigment and a binder. Common fillers are china clays, i.e. silicates. Pigments are often calcium carbonate or titanium dioxide. Binders are often styrene butadiene rubber (SBR) latex. Styrene odours can sometimes arise from these. Other binders used are styrene/maleic anhydride, acrylate copolymers, or alkyd polyesters. Impurities and monomers such as ethyl benzene and butyl acrylate can give rise to odours.

Bleaching the pulp with chlorine dioxide can result in chlorinated phenolics. These are highly tainting and can be chemically converted, for example by bacteria, into even more powerful tainting compounds. A review of taints from paperboard packaging was carried out by Tice and Offen (1994). Odour can be caused by microbial action. Under anaerobic conditions volatile fatty acids (VFAs) can be produced. These VFAs include acetic acid (vinegar) and butyric acid (rancid butter). Wood, from which paperboard pulp is produced, contains glycerides, resin acids, alcohols, waxes and fatty acids. Hydrolysis of the glycerides can occur via lipase enzymatic activity during storage. Autoxidation of unsaturated compounds occurs to form hydroperoxides, which in turn decompose to aldehydes and ketones with hexanal (a boardy mown grass odour) being a common product. Contamination of the paper pulp during paper production can occur. The use of recycled paper carries with it the risk of contamination resulting in taint. The most common causes of taint from paper are chlorocresols, phenols and anisoles. Bromine derivatives of these compounds have also been reported (Whitfield et al., 1989; 1997)

Kraft paper can contain a series of phenol derivatives that are potentially tainting. These include 2-methoxyphenol (guaiacol), 4-methylphenol (p-cresol), 4-vinyl-2-methoxyphenol, 2, 6-dimethylphenol and 4-hydroxy-3-methoxybenzaldehyde (vanillin). All of these compounds have been observed in one sample of paper at Pira International Analytical Laboratory. There is good evidence in the literature that these compounds are all derived from the oxidation of lignin during paper manufacture (Kurschner, 1926). The p-cresol and guaiacol are highly tainting. The odour threshold for guaiacol and p-cresol is 21 and 200 ppb (parts per billion) above water, respectively (Saxby, 1992). Guaiacol can also be produced by biological action on vanillin (Kilcast, 1996).

4.2.4 Paper plastic laminates and metal composites

Paper and plastic laminates are widely used in food packaging. Paper and aluminium foil are used on a large scale in drink cartons and paper tubes for chocolate and crisp packaging. In almost all cases, the foil has an extruded layer of polyethylene on both surfaces to protect the aluminium and provide a heat seal layer. Cardboard tubes are manufactured from spiral wound paper on metal mandrills. The mandrills have to be lubricated and the choice of lubricant is important if odour is to be avoided. If vegetable-based oils are used, aldehydes such as hexanal can be generated from free radical degradation of the vegetable oil. Minute traces of degraded oil containing the aldehydes are sufficient to cause taint in the sealed tubes. Drink cartons are manufactured by extruding a polyethylene layer onto paperboard. If the polyethylene is overheated, thermal oxidation of the polyethylene layer occurs. This results in a taint in the drink. Numerous workers have studied the causes of taints from this type of packaging. Hoff and Jacobsson (1981) discuss the mechanisms involved in the thermal oxidation of polyethylene. They found that out of 44 compounds identified and quantified, fatty acids and aldehydes predominated. Bravo et al. (1992) identified odorous compounds resulting from thermal oxidation of polyethylene. These are predominantly C6 to C9 saturated or unsaturated aldehydes and ketones, the important ones being hexanal, 1-hepten-3-one, 1-octene-3-one, octanal, 1-nonen-3-one, nonanal, trans-2-nonenal and diacetyl (butane 2,3-dione). The effect of processing temperature and time on polyethylene has been investigated (Bravo and Hotchkiss, 1993). Hexanal, a well-known tainting compound is often the most abundant aldehyde produced. α-Unsaturated aldehydes and ketones were found to be responsible for much of the odour associated with thermally oxidised polyethylene. Villberg et al. (1997) identified aldehydes and ketones that were responsible for taints in a range of polyethylenes.

Paper and metal composites are used in packaging of cakes and fish such as salmon where a decorative 'upmarket' presentation is required. A polyethylene or a coating may be applied to the metal as metal can catalyse deterioration reactions with foods. Paper is also used with a mineral coating containing finely divided metal particles applied to the food contact surface. This coating is called a susceptor layer. It becomes very hot and browns the food. The elevated temperatures can greatly accelerate taint forming reactions and migration rates of tainting compounds into the food.

4.2.5 Metal

In most cases the metal is not in direct contact with the food. Direct metal contact can affect the taste and appearance of the food. Coatings are widely used on metal packaging; these are often organosol PVC, epoxyphenolic, polyester or PET. Taints are not that common compared to other types of packaging. This may be due to the stoving processes carried out on the coated metal during can production and retorting of the food in the cans before sealing.

4.2.6 Regenerated cellulose films

Twist wraps for boiled sweets are typical examples of regenerated cellulose film packaging. Taint transfer is rare and when it occurs it is often as a result of poor quality inks. A softener is often used which has the effect of plasticising the film. Softeners used include glycerol and sorbitol. These compounds do not have particularly strong odours or tastes.

4.3 Sources of taints

There are a number of excellent reviews covering taints from packaging. Reineccius (1991) covered many of the most common compounds causing taints. Tice (1996) summarises a great deal of experience in the field of taints from packaging.

4.3.1 External contamination

Contamination from external sources can occur from the use of cleaning fluids and disinfectants, preservatives or fumigants in storage areas. Phenol may be present in wood used in storage areas or pallets. It may originate from coatings or be derived from lignin, a natural constituent of wood. If the wood is washed with sodium hypochlorite solution (domestic bleach) chlorophenols may be generated from the phenol. Wood preservatives used on wood pallets can contain chlorocresols, which can be transferred into polymer pellets, which are subsequently converted into packaging. These examples form a group of contaminants which occur on a regular basis as the cause of taint problems. Often they are characterised by a medicinal taste and an antiseptic lotion or a musty odour. These are the cresols, phenols and anisoles. Chlorophenols and bromophenols give antiseptic lotion odours and the anisoles give musty odours. The anisoles often arise from bacterial methylation of the corresponding phenols. The thresholds for detection of these compounds particularly the anisoles are very low. Examples of odour and taste thresholds are shown in Tables 4.1 and 4.2.

Table 4.1 Odour thresholds above water

Chemical	Odour threshold/ ppm in the water	Reference
2,6-Dichloroanisole	4×10^{-4}	Fazzalari, 1978
2,4,6-Trichloroanisole	3×10^{-8}	Fazzalari, 1978
2,3,4,6-Tetrachloroanisole	4×10^{-6}	Fazzalari, 1978
Pentachloroanisole	4×10^{-3}	Saxby, 1992
Chlorophenol	1.2	Saxby, 1992
2,4-Dichlorophenol	0.2	Saxby, 1992
2,4,6-Trichlorophenol	0.3	Saxby, 1992

Packaging materials as a source of taints 71

Table 4.2 Taste thresholds in water

Chemical	Taste threshold in water/ppm	Reference
2,4,6-Trichloroanisole	2×10^{-5}	Saxby, 1992
2,3,4,6-Tetrachloroanisole	2×10^{-4}	Saxby, 1992
Chlorophenol	6×10^{-3}	Fazzalari, 1978
2, 4-Dichlorophenol	3×10^{-4}	Saxby, 1992
2,4,6-Trichlorophenol	2×10^{-3}	Saxby, 1992

It is vital that the presence of all of these compounds is screened for as a set. The assumption has always to be made that all are present and a specific sensitive analysis with detection limits in the sub parts per billion range must be carried out for all of them. It is very easy to fail to identify the cause of a taint by simply not monitoring for the presence of one of these compounds. Dietz and Traud (1978) list the odour and taste threshold values for a comprehensive list of phenolic compounds. The list includes 126 odour thresholds and 36 taste thresholds in water.

4.3.2 Printing inks and varnishes

Poor quality printing inks are one of the most common causes of taints. The importance of using good quality inks from reputable suppliers cannot be overemphasised. Plastic laminates are printed using either gravure or flexo printing processes. Gravure involves the transfer of the image to the plastic by means of an engraved roller. Flexo transfers the image using a rubber surface mounted on the roller. Solvents used are blends of ethanol and propanol with ethyl, propyl or butyl acetate and higher boiling solvents such as glycol ethers to modify stability on the press. Here press stability refers to solvent evaporation rate. The resins used are typically nitrocellulose and polyurethane or polyamide. It is comparatively rare to have taint problems that are attributable to the resins, provided food packaging grade inks are purchased from reputable ink companies. An adhesion promoter is added to improve adhesion onto polyolefin films. Typical adhesion promoters are titanium acetyl acetonoate, or IA10, a titanium chelate of isopropoxybutyl phosphate. Titanium acetyl acetonate can give rise to taint if a printer adds too much, for example in attempting to improve adhesion. The actual chemical responsible for the taint is pentane-2, 4-dione present in the adhesion promoter, which has an antiseptic medicinal odour and taste.

Over-varnishes are applied to protect the ink from scuffing or the action of greases, and to give extra gloss. In the past these varnishes incorporated drying oils. Free radical and air oxidation resulted in a slow polymerisation and cure. Since the mid-1980s UV free radical curing of varnishes has progressed at such a rapid pace that this process now accounts for most of the varnishes used in food packaging. There is a misconception that still remains that UV-cured varnishes have the inevitable problem of odour. Many of the problems that were initially experienced with UV-curing inks and varnishes have been largely overcome through the introduction of high purity higher molecular weight prepolymers and high purity and more efficient photoinitiator systems. Reactive diluents are now used rather than simple solvents. The diluent itself takes part in the polymerisation process and is incorporated into the polymer bulk. Unlike a solvent, it is then unable to migrate into the food. In food packaging UV-cured coating is now a mature technology with a sound underlying scientific and technical understanding of the processes involved. Provided that high quality inks and varnishes are used, that the application requirements are followed and good working practises are maintained, taint issues rarely occur. The problem has been that printers have seen the technology as enabling them to print at very high speeds and ship the packaging without sufficiently airing the printed material. The airing process allows a large proportion of any residual volatile compounds to evaporate from the food packaging. A taint may occur if the minimum period of 3 days turning and airing recommended by ink manufacturers is not observed. Usually taint problems can be traced back to poor work practice or the use of cheap poor-quality inks or solvents.

It is important that the UV lamps used to cure coatings are checked to ensure proper light emission. Line speeds should be correct for the application. If the line speed is too slow, rapid cure occurs on the outer surface and an under-cured inner layer results. Migration is then backwards in towards the food, because the outer surface has become a hard cross-linked solid. If the viscosity of the ink is incorrect, unwanted side reactions occur during the curing process. If the resin matrix becomes too viscous too quickly the photoinitiator radical that initiates the polymerisation process is not able to migrate into the bulk of the resin and cannot carry out its function properly. Photoinitiators that are activated by the UV light react with intermediates and each other to give 'cage products'. Methyl benzoate, which has a strong pungent odour, is one such example. Often more than one photoinitiator is used and there is now a wide range of photoinitiators tailored to suit different applications. Resins are very often based on acrylate type prepolymers. The resins are rarely the cause of taints. An explanation of UV radiation curing technology (radcure) is provided by Roffey (1982) and Crivello and Dietliker (1998).

Litho printing is still widely used for printing of carton board. In this printing process, a petroleum-based distillate solvent is used. The image is created by separating the ink from regions where it is not wanted on the image using a water-based fount solution. For food packaging it is important that the solvent does not contain aromatic distillates such as toluene or xylene, which contribute considerably to odour. The pore structure of the board can affect solvent penetration into the board. A more open pore structure resulting from a lower degree of calendering during manufacture can affect the solvent absorption characteristics of the board.

4.3.3 Adhesives

Adhesives were often solvent based with the solvent softening or dissolving the polymer to be bonded, or providing a resin solution which dries to form the bond. Increasingly adhesives for food packaging are of the solvent-free type. These are often polyester or polyurethane. Typically, one or two part systems are used with a polyol and an isocyanate hardener. A range of ester and ether urethanes is available. The isocyanate is often a prepolymer. Urethanes are widely used to bond polyolefins to PET. If the polyol is not of adequate purity, reaction of alcohols or diols with aldehydes or ketones can occur. 4,4,6-Trimethyl-1,3-dioxane formation is an example of a class of reaction where an alcohol or a diol can react with a carbonyl to form ketals and acetals or cyclic ketals and cyclic acetals. McGorrin *et al.* (1987) describe the identification of this tainting compound. The odours of the cyclic compounds vary, being described as sweet camphor, honey and musty. These compounds have low odour thresholds, typically in the sub parts per billion range. They are very difficult to identify by GC-MS owing to the fact that fragmentation occurs in the mass spectrometer to give relatively low mass fragments and no molecular ions. They can be present in the packaging as a result of solvent/adhesive interaction or as impurities in poor quality solvents. Sanders (1983) describes the formation of dioxanes. The compound 2-ethyl-5,5-dimethyl-1,3-dioxane was found to be the cause of a taint in water (Preti *et al.* 1993). The odour threshold for this compound above water was found to be 0.01 ppb (w/v).

Cold seals are pressure-sensitive adhesives widely used on sweet wrappers such as chocolate bars. These adhesives are often made from styrene butadiene acrylate copolymer latex. Residual ethyl benzene, styrene and butyl and ethyl acrylate are common odorous compounds, the presence of which has to be controlled.

4.3.4 Additives

Although polymer additives can give rise to taint, such incidents are very rare. The common additives used in food packaging are generally not highly tainting chemical compounds. When there is taint associated with an additive, it usually arises from the presence of impurities or degradation products in the additive. Poor quality or degraded slip agents such as stearates or oleates can give rise to rancid odours and soapy tastes. Some grades of polypropylene have clarifying agents added such as those of the benzylidene sorbitol type. Clarifying agents work by reducing the chain lengths of the polymer and hence its ability to scatter light. Impurities that can arise in poor quality benzylidene sorbitol clarifying agents are often methyl-substituted benzaldehydes which have almond odours.

4.3.5 Alteration in the packaging format

Alteration in the packaging format usually results from pressure for a new presentation from those responsible for the marketing of the product. A supermarket may require a particular packaging style format. A promotional item may be included, for example a decoration set, candles, ribbons and so on. Taints can arise from these promotional items. Taints can also arise from the cartons. If the cake is packed hot into the box, volatiles can be generated from the carton or base tray. These may become trapped in the box if an overwrap is used. When the cake is subsequently placed in cold storage these volatiles condense on the outer surface of the cake. Icings can be particularly sensitive to tainting compounds.

Packaging drinks in HDPE bottles with shrink-wrap labels rather than in cartons or more expensive PET bottles provides another example. The development of thin films that can be printed and then shrunk over the outer surface of drink bottles has resulted in new more appealing packaging. The residual solvent content in the sleeves should be monitored and kept as low as possible. Solvents that are present in the sleeves tend to migrate through the HDPE (which is a poor barrier) into the drink rather than be lost to the air through the relatively high barrier material of the sleeve. In general, solvents such as butyl acetate, glycol ethers and tetrahydrofuran (which is sometimes used to weld the seam on the sleeves) can give rise to taint problems in some drinks such as chocolate-flavoured milks.

A common source of taints is promotional printed items in packaged foods. These may be introduced in a promotion or marketing campaign when time restrictions prevent consideration of the likely taint risks. These take the form of printed cards or toys that are flow wrapped in a transparent film. Promotional items such as scratch-off cards are often printed with styrene rubber lattices and petroleum-based inks. Printers of such material may not be familiar with the requirements of the food packaging industry and they may not be using low odour inks of high purity and quality. Toys may be obtained from the Far East where raw material quality may not always be as good as in Europe. Such items may be screen printed with the result that relatively high boiling solvents such as isophorone are used. The choice of flow wrap material is important and checks should be made to ensure that where sensitive foods such as cakes are used, the flow wrap is an effective barrier. Polypropylene is commonly used to flow wrap the toy. The polypropylene should be of sufficient gauge to resist puncture and should be barrier coated with a polyvinylidene chloride (PVDC) barrier layer if the promotional item contains significant amounts of solvent or other tainting substances.

Packaging materials as a source of taints 75

Table 4.3 Types of packaging material producing radiolysis products

Polymer	Radiolysis product
Polystyrene	Acetophenone, benzaldehyde, phenol, 1-phenylethanol, phenylacetaldehyde
Polyamide	Pentamide and traces of amides, formamide, acetamide etc.
Polyvinyl chloride	Octane, 1-octene, acetic acid 2-ethylhexylester 1-octanol
Polyethylene	Very low concentrations of hydrocarbon, aldehyde, ketones and carboxylic acids, disappear rapidly on storage, carboxylic acids still detectable after 6 weeks
Polypropylene	Antioxidant decomposition products and aldehydes, ketones and carboxylic acids

4.3.6 Packaging sterilisation by irradiation

Sterilisation by irradiation of packaging can generate potentially tainting compounds as a result of degradation of some polymer types. Generation of potentially tainting compounds from a range of polymers has been studied by Buchalla *et al*. (1993, 1998). Single doses of ca. 25 kGy are sufficient to produce detectable amounts of low molecular weight radiolysis products. The types of product are listed in Table 4.3.

It can be seen from this table that vigilance is required, since potentially tainting compounds are formed particularly in the case of polystyrene. Kilcast (1990) warns of tainting observed with polystyrene and PVC after irradiation with 2.6–3.9 kGy. A review by Buchalla *et al*. (1993) covers in detail the issues of taint arising from irradiated polymers.

4.4 Chemicals responsible for taints

The chemicals listed in the Table 4.4 have been found to cause taint. These compounds have all been associated with taint produced by paper, polymer films, coatings or resins.

Aldehydes and ketones are two classes of compounds that are often responsible for taints. These include the conjugated unsaturated carbonyl compounds. These compounds are particularly odorous. An excellent evaluation of the importance of these compounds was made by Koszinowski and Piringer (1986). These workers synthesised a range of unsaturated carbonyls and determined the odour thresholds as a function of carbon chain length. They found that the most odorous compounds were those with eight or nine carbon atoms. There are numerous possible reactions that can occur with components in packaging materials that result in the formation of conjugated unsaturated ketones and aldehydes. Isolated double bonds will, if possible, undergo chemical transformation and move into conjugated systems. Once produced, the α- and β-unsaturated compounds are

Table 4.4 Common tainting compounds

Compound	Taint	Source
1,1-Diethoxyethane	Jasmine odour	Impurity in ethyl acetate printing ink solvent
2-Ethyl-5,5-dimethyl-1,3-dioxane	Sweet, nutty, woody	Reaction of 2-propenal with neopentyl glycol
3-Isopropyl-2-methoxypyrazine	Musty odour	Suspected formation involving unknown bacteria
4,4,6-Trimethyl-1,3-dioxane	Musty odour	Reaction of paraformaldehyde with 2-methyl-2,4-pentanediol in a coating
2,2,6-Trimethyl-1,5-dioxane	Sweet, camphor odour	Reaction of acetone with 1,3-butanediol
2,2,4,5-Tetramethyl-1,3-dioxane	Camphor, liniment odour	Reaction of acetone with 2,3-butanediol
2-Ethenyl-2,5-dimethyl-1,3-dioxane	Musty, liniment odour	Reaction of methyl ethyl ketone with 1,2-propanediol
4-Methyl-4-mercaptopentan-2-one	Catty urine odour	Hydrogen sulfide adduct of mesityl oxide from diacetone alcohol with hydrogen sulfide from meats
4-Phenylcyclohexene	Synthetic latex odour	Diels Alder condensation product of styrene with butadiene in binder coated paper
Acetaldehyde	Pear-like odour taste	Degradation product sometimes formed during processing of PET
Benzophenone	Geranium odour	Photo-initiator in UV inks and varnishes
Aliphatic acids	Short chain lengths particularly odorous, e.g. butyric acid has a rancid off odour	Recycled paper Degraded lubricating oil
Alkyl acetates e.g.: ethyl acetate propyl acetate butyl acetate	Fruity odour	Flexo and gravure print solvent
Alkyl substituted benzenes	Hydrocarbon	Styrene/butadiene latex binder
α-Methyl styrene	Hydrocarbon plastic	Styrene/butadiene latex binder, also present in some EVOH grades
Butyl acetate	Pear drop odour, fruity taste	Printing inks
Chlorocresol	Medicinal odour and taste	Disinfectants
Cumene (isopropyl benzene)	Hydrocarbon	Styrene/butadiene latex binder, also present at trace level in some EVOH grades

Table 4.4 *Continued*

Compound	Taint	Source
Cyclohexanone	Sweet pungent odour	Screen-printing solvent
Di/tribromophenol	Medicinal taint	Bromination of phenol, an impurity in polymers, used as wood preservers on pallets
Di/trichlorophenol	Medicinal taint	Wood preservers, herbicides, industrial chlorination of natural or synthetic phenols
Dichlorobenzene	Medicinal taste	Disinfectant, drain cleaner, fumigants
Diphenyl sulfide	Cabbage-like odour	Photoinitiator for cationic inks
Glycol ethers e.g. 2-butoxyethanol 2-ethoxypropanol	Soapy taste	Printing solvent
Guaiacol	Smoky phenolic	Lignin derived in kraft paper
Hexanal	Board/mown grass odour	Lipid degradation commonly associated with paper
Isophorone	Pungent brown sugar odour	Screen printing inks common cause of taint from promotion toys
Methyl benzaldehyde	Almond odour	Impurity in clarifying agent for polypropyelene sheet and containers
Methyl benzoate	Pungent herbal odour	UV ink/varnish unwanted side reaction
n-Propyl benzene	Hydrocarbon	Styrene/butadiene latex binder
Naphthalene	Petroleum odour/taste	Litho print solvent
p-Cresol	Phenolic	Lignin derived in kraft paper
Pentan-1,2-dione	Medicinal, chemical taint	Present in titanium acetyl acetonoate adhesion promoters
Propylene glycol monobutylether	Chemical taste	Printing inks
Styrene	DIY fibre glass car repair odour	Monomer in polystyrene and wide range of coatings Migration greater from high impact polystyrene (HIPS) than crystal PS
Thioglycollic acid alkyl esters	Pungent strong stale beer	From alkyl tin stabilisers used in PVC
Toluene	Petroleum odour/taste	Litho print solvent
Tribromoanisoles	Musty odour	Bacterial methylation of corresponding phenols
Trichloroanisole	Musty odour	Bacterial methylation of corresponding phenols
Trimethylanisole	Musty odour	Contaminant in rubber seals

chemically very stable owing to the stabilising effect of the conjugation. Such compounds tend to persist in the packaging because of a comparatively low volatility and moderate polarity. The low taint thresholds make the identification of these compounds difficult and sometimes impossible without selective concentration procedures. This class of compounds can produce serious taint problems in food packaging. Aliphatic alcohols and alkanes are rarely the cause of taints. Relatively high concentrations are required before they can be smelt or tasted.

4.5 Main foodstuffs affected

The most frequently affected foodstuffs are confectionery and drinks. Confectionery is a sensitive food to taint. Wrappers are often laminates with high print coverage. Packaging surface area is relatively large compared to the bulk of the food. The flavour of chocolate is easily affected by chemicals. Fats present in the chocolate are in direct contact with the wrapper and partitioning of the tainting compounds occurs onto the surface of the chocolate. The outer surface of the food can therefore act to concentrate the tainting compound in a small amount of food. Duek-Jun and Halek (1995) found that the extent to which solvents partition into the chocolate depends upon the temperature, the degree of crystallinity of the chocolate and the fat content. The order of partition was toluene > isopropanol > methyl ethyl ketone > ethyl acetate > hexane.

Table 4.5 shows the relative proportions of foodstuffs affected by taint, which were investigated over a ten-year period at Pira International Analytical Laboratory.

Table 4.5 Relative proportions of foodstuffs affected by taints

Foodstuff	% of cases
Meat	3
Bread/pizza	8
Soup	5
Flavoured milk drinks	8
Alcoholic drinks	5
Sweets	33
Cakes and biscuits	17
Cheese	3
Crisps	5
Water	5
Beverages	8

Table 4.6 Relative proportions of compounds causing taints

Contaminant chemical type	% of cases
Solvent or ink component	29
Phenols or halogenated phenols/ anisoles	18
Aliphatic aldehydes and ketones	24
Dichlorobenzene	5
Styrene	11
Others[a]	13

[a] Others refers to compounds that do not re-occur as the source of taints.

Table 4.6 shows the proportion of cases investigated over the same ten-year period at Pira International Analytical Laboratory associated with a particular chemical class of contaminant.

Case study data from Nestlé 'were reviewed by Huber et al. (2002). Huber reported styrene accounting for 15% of cases, halogenated phenols 15% and solvent 28%.

In recent years the Pira International Analytical Laboratory has begun to identify bromoanisoles as the cause of musty odour and chemical taste in packaging where previously it was due to the presence of chloroansioles. In such cases the corresponding bromophenols, particularly 2,4,6-tribromophenol are also present. Until recently the source of these compounds has remained a mystery apart from studies reported in the literature. An excellent paper by Whitfield et al. (1997) traced the source to wood preservers used on wooden pallets. Whitfield et al. warned of the risk posed by the use of phenol-based wood preservers. At the time of going to press a large investigation carried out by Pira has established that 2,4,6-tribromoanisole derived from 2,4,6-tribromophenol was the cause of a musty odour reported simultaneously by at least two packaging manufacturers. The problem was traced back to one particular supplier who had used wooden pallets treated with 2,4,6-tribromophenol (as the phenate, i.e. alkali salt form) to transport polyethylene raw material by sea. This case highlights a trend observed by the author of an increasing frequency of bromoanisole contamination in packaging. Wooden pallets used to export product have to be treated to prevent spread of tree and plant diseases. Until phenol-based wood preservers are discontinued, they will continue to be one of the most common, costly and easily preventable packaging-related causes of taint, and wooden pallets will remain high risk contaminating surfaces for packaging materials.

4.6 Instrumental analysis of taints

The technique of choice is GC-MS. This technique has advanced steadily since the early 1990s. Sensitivity and spectral quality have improved dramatically. Electron impact mass spectral libraries have greatly increased in size. Advances in capillary gas chromatography columns with low bleed and high resolution have provided the ability to separate relatively easily a large number of chemical compounds. These compounds can then be reliably identified by computer comparison against library spectra. Identifications should still be confirmed by running a pure standard and also by comparing retention times and mass spectra. Occasionally the libraries are wrong or the spectrum obtained on the instrument is different to that in the library because it may have been obtained on a different type of instrument.

A choice of instrument types is available: quadrapole or ion trap. Mass spectra obtained using different mass spectrometers are not always identical for a particular compound. Ion traps are more sensitive than quadrapoles and there is an ability to carry out MS-MS operation and observe the fragmentation of daughter ions. Chemical ionisation (CI) is a technique where a molecular ion can be obtained from easily fragmentable molecules such as ethers by the gentle ionisation obtained by using chemical reagent gases such as ammonia or methane. It is useful for compounds that easily fragment where electron impact ionisation does not provide molecular ions or unique characteristic spectra.

Infrared (IR) scanning detectors are also available. The detector is non-destructive and GC-MS-IR instruments can be purchased where the sample passes through the infrared detector first before passing into the mass spectrometer. This provides an additional means of identifying compounds. Unfortunately the gas phase IR spectra obtained are different from the commercially available spectral libraries, with the result that suitable reference libraries are limited. In addition, the sensitivity of the detector is not usually adequate. The IR detector also adds considerably to the cost of the equipment.

If odours are to be investigated, it is essential that the equipment be fitted with an odour port. This involves the use of a split at the exit of the analytical column before the entrance to the mass spectrometer. This is best achieved by the use of a zero volume T piece, situated inside the gas chromatograph oven. One arm of the T piece is connected to the odour port and the other is connected to the mass spectrometer. A simple design is described in the literature (Acree *et al.*, 1976). The art of GC olfactometry has been refined over the years (Acree *et al.*, 1984; 1997). The odour port should consist of a heated deactivated silica connection to the T piece. Humidified air is mixed with the gas eluting from the analytical column and presented to the human nose in a glass pocket-shaped vessel. The balance of gas flow is an important consideration. Insufficient flow to the mass spectrometer results in reduced sensitivity from the mass spectrometer. Too high a flow rate to the mass spectrometer and not enough to the odour port results in no odours being smelt. Although it is possible to install a variable splitter valve to adjust the split ratios, this can result in peak broadening and reduced peak resolution. The easiest approach is to allow the mass spectrometer to have its optimum flow (typically $1\,\mathrm{ml\,min^{-1}}$) and split the rest of the flow to the odour port. This is achieved using a length of deactivated silica tubing as a restrictor into the mass spectrometer (1.4 m for a 0.150 mm ID (internal diameter) restrictor). The odour port is connected by means of wide bore tubing. Alternatively a 50/50 split can be set up by connecting equal lengths of restrictor to the mass spectrometer and odour port. A make up gas can be introduced at the split to increase the velocity of the flow to the nose port. This can enable simultaneous detection at the odour port and mass spectrometer. This equipment can all be purchased off the shelf from laboratory equipment suppliers.

The use of an odour port greatly assists an odour investigation. This is another reason why odour investigations are usually easier than taste problems. Identification of the cause of a taste requires the food to be fortified with the suspected tainting compounds. Tasting chemicals in this manner raises issues of safety. The only other solution is comparison of concentrations to taste thresholds in the literature. Unfortunately there is a dearth of comprehensive taint threshold data.

Figures 4.1 and 4.2 illustrate the basic steps in investigating undesirable tastes and odours. The process starts with a sensory panel test by experienced panellists to establish a reliable description of the taste. An accurate description is invaluable in choosing subsequent lines of investigation. Samples of non-odorous and odorous packaging are then prepared using such techniques as steam distillation (a standard technique here is the Lickens–Nickerson technique or solid phase microextraction (SPME). Preparation techniques are discussed in section 4.7. They can then be analysed by GC-MS to produce chromatograms which can be compared in order to isolate a tainting compound.

4.7 Sample preparation techniques

The sample preparation stage in an investigation is usually critical to the usefulness of the results obtained. There are a number of sample preparation techniques.

4.7.1 Dynamic headspace

The dynamic headspace technique is the one most commonly used in the author's laboratory. It offers the greatest sensitivity of all the sample preparation techniques. Since odours arise at room temperature, the chemicals responsible for the complaint odour will have sufficient volatility to be

82 Taints and off-flavours in food

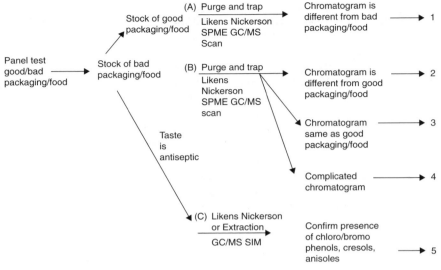

Key

Follow route (A) first to avoid contamination of equipment. Route (C) is a special case where extraction GC/MS SIM is needed to detect the tainting compounds

1 Continue with analysis of bad packaging/food
2 Identify all compounds not in good packaging/food. Analyse tainting compound compare mass spectra, retention times, taste threshold, taste description
3 Analysis insufficiently sensitive. Quantify packaging. Identify compounds present only in the bad packaging. Monitor for these compounds in the good/bad food in SIM mode
4 Use food simulants e.g. water/ethanol, lactose in contact with bad/good packaging with panel tests until a taint has been transferred. Analyse simulants. Monitor in SIM mode for the presence of suspect compounds in the good/bad food
5 Compare concentration to experimental/literature taste threshold

Fig. 4.1 Identification of the chemicals responsible for an undesirable taste.

amenable to this technique. Depending on the form of the samples, an inert stripping gas such as helium is passed over or through the sample, which is maintained at room temperature. The stripped volatiles, including the odorous chemicals, are trapped onto an adsorbent trap. This trap is a tube usually containing Tenax TA, a modified polyphenylene oxide polymeric material that adsorbs a wide range of compounds. Brown (1996) compared the performance of a range of adsorbents. The best general-purpose adsorbent for taint investigations is Tenax TA.

A thermal desorber is used to transfer the volatiles from the trap to the gas chromatograph. In this apparatus the trap is heated to temperatures up to 200°C and the volatiles desorbed from the trap are carried, using the gas chromatograph carrier gas stream, onto a short length of deactivated fused silica tubing. This tubing is maintained at approximately −150°C by means

Packaging materials as a source of taints 83

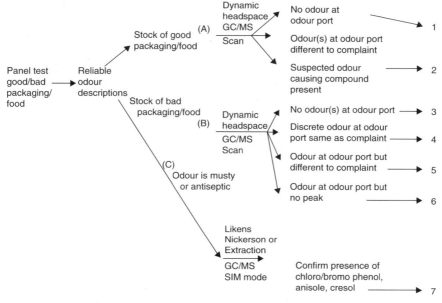

Key

Follow route (A) first to avoid contamination of equipment.
Route (C) is a special case where extraction GC/MS SIM
is needed to detect the odour causing compounds.

1 Continue with analysis of bad packaging/food
2 Odour cause not proven unless concentration in good samples is less than odour threshold and concentration in bad samples is greater than odour threshold
3 Odour causing chemical may not be amenable to GC/MS e.g. aliphatic amines, chlorophenols
4 Identify all compounds in the odour region of the chromatogram and differences between good and bad samples. Inject suspect odorous compounds; compare mass spectra, retention times, odour at odour port
5 The complaint odour may be a combination of two or more individual odours
6 GC/MS may not be sufficiently sensitive. Analyse concentrated extracts of bad packaging. Screen for likely compounds in SIM mode. Examine the chromatograms for possible precursor chemicals that may provide a clue
7 Compare concentration to experimental/literature threshold

Fig. 4.2 Investigation of an odour problem.

of a stream of liquid nitrogen. The desorbed compounds are focused as a narrow band in the tubing. Once all the compounds have been desorbed from the trap, the deactivated silica tubing is rapidly heated to 200°C and the trapped compounds volatilised into the GC analytical capillary column. The equipment incorporates a split before the top of the cold trap so that a proportion of the sample can, if required, be split out to waste. This is a useful feature to ensure that sample overloading does not occur.

Tenax can decompose after prolonged heating and aromatic artefacts arising from the trap can appear. Sample breakthrough can also occur, so that excessive purging of the helium carrier gas through the trap can cause the compounds of interest to be lost. Activated carbon is used for very volatile compounds. Trap mixtures of several adsorbent materials are available to ensure that all compounds of interest are trapped. A data sheet is available listing suitable trap packing materials, operating temperatures and advice on packing of multiple sorbent traps (Perkin Elmer, 1993).

If there is a choice, it is better to use an overpressure of stripping gas rather than a vacuum to drive the volatiles from bags or containers under investigation into the trap. The reason for this is a lower background achieved using an overpressure of carrier gas since laboratory air is excluded. If application of an overpressure is not possible, the air may be sucked out through the trap using a vacuum pump. A blank should be run at the same time to ensure that volatiles in the laboratory air are accounted for. If the sample is in the form of granules, powder or liquid these are best placed in a glass chamber. The chamber is then purged with the stripping gas and the volatiles collected on the trap. If the sample is in sheet form and only one surface is to be sampled, the sheet can be sealed in a glass cell with the surface of interest exposed to the stripping gas.

Commercial purge and trap equipment is widely available. Purging of the sample, trapping and injection of the volatiles from the trap are all carried out automatically. The principle of this equipment is described by (Werkhoff and Bretschneider, 1987) and Badings and De Jong (1985). The equipment enables very sensitive analysis to be carried out. Although usually used for liquid samples, solids such as plastics and papers can also be sampled using commercially available purge and trap instruments. The equipment is very sensitive and contamination of the equipment is always a serious risk. Sample carry-over can also be a problem. The equipment includes a moisture control system (MCS) line, which is a length of tubing usually maintained at low temperature to condense and remove water. Packaging such as paper can contain appreciable amounts of water. It is seldom advantageous to heat samples, particularly aqueous samples, above 70°C. The presence of water can cause sample carry-over problems from the MCS line, or excessive ion source pressure in the mass spectrometer. A choice is available with commercial equipment. Some instruments trap the purged volatiles on a cryogenically cooled silica capillary trap. Other instruments trap the volatiles onto a Tenax trap. The advantage of the latter is that the Tenax can be dried by passing carrier gas through the trap. There is also less risk of losing high boiling and water-soluble compounds. Purge and trap equipment offers better reproducibility and automation. Off-line dynamic headspace sampling provides versatility of sampling and avoids contamination of the GC and purge and trap equipment until more is known about the concentration of volatiles in the sample.

4.7.2 Static headspace
The static headspace technique involves heating the sample of interest in a glass vial in an auto sampler for a period of time. Volatile compounds partition into the headspace in the vial. A portion of the headspace is then injected for analysis by gas chromatography. As a general rule dynamic headspace sampling is more sensitive than static headspace sampling. For this reason dynamic headspace sampling is preferable for the identification of tainting compounds. Once the identity of the tainting compound has been established, static headspace can then be used to screen for the presence of the compound in a range of samples. The mass spectrometer may be operated in the selected ion-monitoring (SIM) mode, which offers approximately one order of magnitude increase in sensitivity. Static headspace offers the advantage over dynamic headspace sampling in being more reproducible and faster to carry out. Static headspace can be used to quantify the tainting compound to enable comparisons to be made with taste and odour thresholds.

4.7.3 Solid phase microextraction (SPME)
Solid phase microextraction is a relatively new and useful technique whereby a stationary phase similar to those used on GC columns is coated onto a glass fibre. The fibre is placed in or above the sample. The compounds of interest partition into the stationary phase. The fibre is then placed into the hot injection port of the gas chromatograph and the adsorbed volatiles are thermally desorbed onto the GC analytical column. The partitioning time required for the fibres is usually quite short, typically 10 min. A range of fibres is available covering the polarity range. Polymethyl disiloxane (PMDS) phases are used for non-polar compounds and acrylate polymers for the polar compounds.

The sensitivity of the technique is not as good as dynamic headspace. This is obvious when one considers that SPME is an equilibrium process, whereas dynamic headspace often involves trapping all of the stripped volatiles. It is favoured by some laboratories equipped with ion trap mass spectrometers. They are able to trade the sensitivity of the ion trap for sample preparation convenience. The technique is described in a series of applications notes (Supleco, 1994; 1995; 1996; 1997). The range of compounds for which the technique is applicable has been extended by 'on fibre derivatisation' procedures, for example for organic acids (Clark and Bunch, 1997).

4.7.4 Steam distillation
The most widely used apparatus is that designed by Likens and Nickerson (Likens and Nickerson, 1964). This is a solvent codistillation technique where the sample (A in Fig. 4.3) is boiled with water. The steam drives the

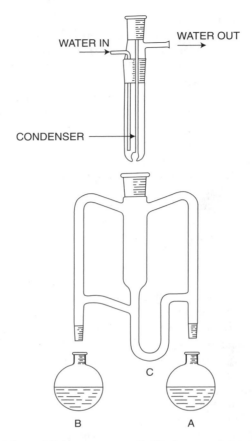

Fig. 4.3 Likens–Nickerson steam distillation extraction apparatus.

volatiles from the sample and the volatiles are then condensed into a water-immiscible volatile solvent such as diethyl ether or pentane. The volatile compounds from the sample partition into the organic solvent (point C), and the organic solvent is returned to a separate heated reservoir (B in the diagram). The compounds of interest are gradually concentrated in the organic solvent by this process. Consideration must be given to the pH of the sample in the water. Basic compounds will be present as water-soluble involatile salts in boiling water at low pH, and acids as the corresponding salts in boiling water at high pH. The procedure is therefore carried out under both basic and acidic conditions. The bulk of the water immiscible solvent is then removed by evaporation.

Extreme care is required during this evaporation stage to ensure that volatile compounds are not lost. The Kurderna-Danish evaporation apparatus (Fig. 4.4) is widely used.

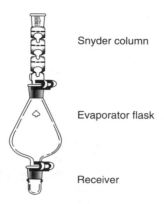

Fig. 4.4 Kurderna-Danish evaporation apparatus.

The apparatus consists of a glass receiver containing the solvent sample to be evaporated. The receiver is attached to an evaporator flask. The top of the flask is attached in turn to a glass Snyder column through which the solvent vapour can escape by pushing up a series of glass flaps. Condensed volatiles can trickle back into the bottom receiver. The apparatus is operated with the receiver immersed in a steam bath. It is vital that the solvent in the receiver is not evaporated to dryness. The solvent is maintained at a steady boil in the receiver allowing recondensing solvent continually to flow back down the sides of the flask. The apparatus should be washed down with a little solvent and the washing combined with the concentrated extract. The highest analyte recoveries are obtained by adding a small aliquot (e.g. 1 ml) of a high boiling keeper solvent in which the analyte of interest is fully soluble at the start of the concentration procedure. Optimised Likens–Nickerson extractions have been described (Bouseta and Collin, 1995). These method variations aim to avoid or minimise the degradation changes that can be induced in sample matrices during the steam distillation.

Steam distillation is particularly suitable for the analysis of food samples where intractable matrices are involved. For packaging analysis there are disadvantages and these include:

- generation of compounds by the boiling process
- complicated chromatograms with many peaks
- contamination risks from all the glassware used
- very time consuming
- risk of loss of tainting compounds of interest during evaporation.

4.7.5 Liquid extractions

Liquid extraction is seldom a successful approach. Often the solvent can extract a large number of compounds giving a complicated chromatogram.

An exception is illustrated by a case investigated at Pira International Analytical Laboratory. This is described in Section 4.9.1 below. In this case a selective solvent was used to extract a specific class of chemical compound suspected to be the cause of the taint.

4.8 Sampling strategy

It is important to sample in the most meaningful manner in relation to the problem. If there is a complaint of an odour from a crisp bag, for example, excessive heating above room temperature may generate compounds that were not initially present in the tainted sample. Sampling the whole bag including the outer surface may result in the detection of compounds not actually present inside the bag. If there is a complaint of a taint from a polyethylene foil paper laminate, there is no point in carrying out a steam distillation on the whole laminate. The contribution from the paper is irrelevant to the problem, as the foil will be a complete barrier. The compounds on the paper surface, coatings and ink will simply confuse the issue. In addition, steam distillation can generate aliphatic aldehydes and acids from lipids present in the paper, which may then complicate interpretation of the results. Once the tainting compound has been identified on the food contact surface, its origin can be established. It may then transpire that it has been transferred from the outer surface by reverse side contact when the film was in reel form or when the paperboard was stacked.

If the problem is one of taste from a packaging material, it is not always the best approach to look immediately for differences between good and bad samples by GC-MS analysis of the packaging. It is sometimes better to start with taste testing experiments with one or two assessors known to have a sensitive pallet. The assessors may enable the best samples to be selected. It may then be possible to transfer the taint from the packaging into a suitable food simulant which is more suitable for analysis than the food itself. These food simulants can be ethanol or water or dextrose. If the assessors confirm the taint has transferred to the food simulant, the tainting compound has then been isolated from the packaging. This is always a considerable step towards identifying the compound. Controls and complaint samples should always be treated in a similar fashion otherwise incorrect conclusions may be made. For example, cutting a polystyrene film into many pieces whilst leaving a control untainted sample in one piece, may result in styrene monomer migrating to a greater extent from the cut edges of the tainted sample during exposure to solvent or food simulant.

4.9 Examples of taint investigations

This section summarises a number of taint investigations carried out at Pira:

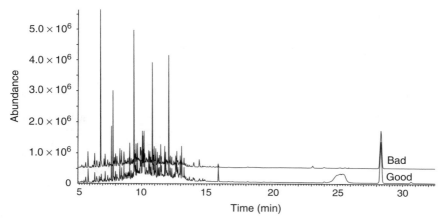

Fig. 4.5 Chromatograms obtained from Likens–Nickerson extract of non-odorous (good) and odorous (bad) bags.

- Case study 1: bread bags
- Case study 2: bags of boiled sweets
- Case study 3: packaged cheese.

4.9.1 Case study 1: bread bags

A complaint of an antiseptic odour on bread bags was investigated. Chlorophenols were immediately suspected. Initially a Likens–Nickerson extraction was carried out. Figure 4.5 shows the chromatograms obtained from good (non-odorous) and bad (odorous) bags. The large number of peaks produced a complicated set of chromatograms. None of the compounds detected had the complaint odour.

The bags were turned inside out and heat sealed across the open end to enable the outer surface to be washed. Sodium hydroxide, 0.1 M, was used as the solvent. The resulting solution was extracted with pentane and diethyl ether and the ether and pentane extracts discarded. This removed alkanes and other compounds of no interest from the bags. The sodium hydroxide washings were neutralised and subjected to a derivatisation procedure using potassium carbonate and acetic anhydride. The acyl derivatives were then extracted from the aqueous phase and injected for analysis by GC-MS (Fig. 4.6).

The procedure was rapid and resulted in a clean extract. The chromatogram from an extract from a non-odorous bag was similar except for a peak at ca. 12 min (Fig. 4.7). Comparison of the mass spectrum of this peak against the nearest library match (Fig. 4.8) identified the compound eluting

90   Taints and off-flavours in food

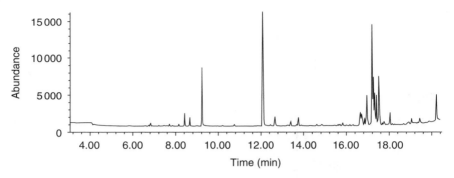

Fig. 4.6   Chromatogram obtained from extract from an odorous bag.

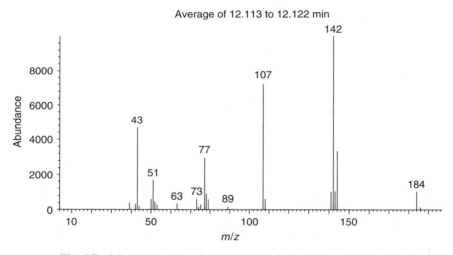

Fig. 4.7   Mass spectrum of the compound eluting at ca. 12 min.

at this point as the acetyl derivative of 6-chloro-*ortho*-cresol (Fig. 4.9). It is important to note that in this example the initial suspicion on the compound identity was wrong. The analysis was able specifically to identify the odour compound as 6-chloro-*ortho*-cresol. The origin of the compound was not established.

4.9.2   Case study 2: bags of boiled sweets
Complaints were received concerning an odour in bags of boiled sweets. The bag was reverse-side printed and laminated. The odour was described

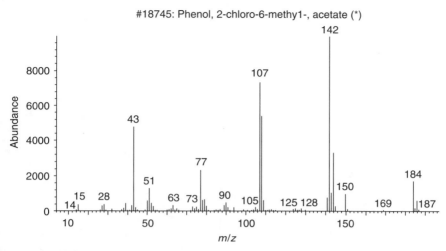

Fig. 4.8 Reference library mass spectrum of acetyl derivative of 6-chloro-*ortho*-cresol.

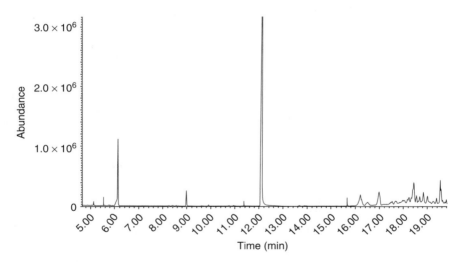

Fig. 4.9 Chromatogram of reference 6-chloro-*ortho*-cresol acetyl derivative.

as jasmine-like, herbal or floral. The odour was related to a batch from one particular printer. Odorous and non-odorous bags were supplied for the investigation. The bags were sampled using dynamic headspace and GC-MS chromatograms were obtained. Figure 4.10 shows a chromatogram from a non-odorous bag and Fig. 4.11 a chromatogram from an odorous bag.

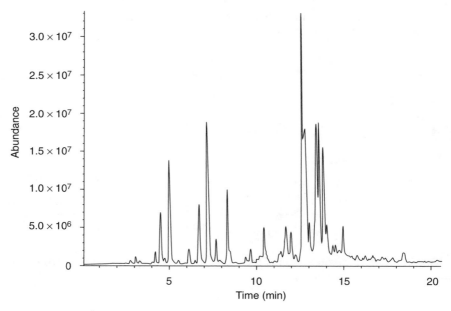

Fig. 4.10 Chromatogram from a good (non-odorous) bag.

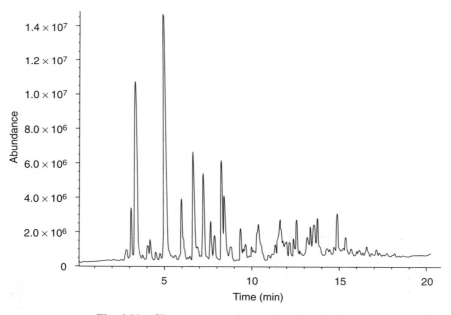

Fig. 4.11 Chromatogram from an odorous bag.

Sniffing at the odour port revealed the presence of the odour in the odorous bag in the region 7 to 8 min on the chromatogram. The scale on the chromatograms is expanded in Figs 4.12 and 4.13.

The chromatograms obtained from the odorous and non-odorous bags both had a peak at 7.2 min. However a co-eluting peak was just visible in the chromatogram from the odorous bag at 7.2 min, which was not present in the chromatogram from the non-odorous bag. The mass spectra obtained

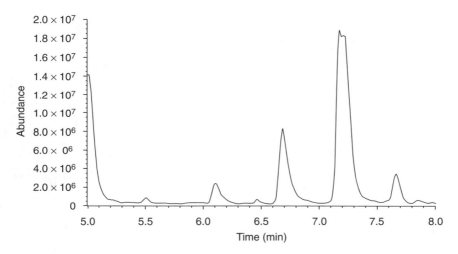

Fig. 4.12 Chromatogram from a non-odorous bag: expanded scale.

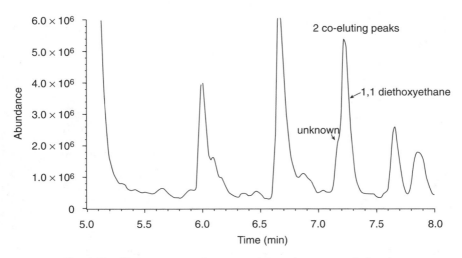

Fig. 4.13 Chromatogram from an odorous bag: expanded scale.

94   Taints and off-flavours in food

from this region of the chromatograms were different, as shown in Figs 4.14 and 4.15.

Comparison of the mass spectrum from the co-eluting peak on the chromatogram obtained from the odorous bag (Fig. 4.15), with the reference library of mass spectra, identified the chemical as 1,1-diethoxyethane. The

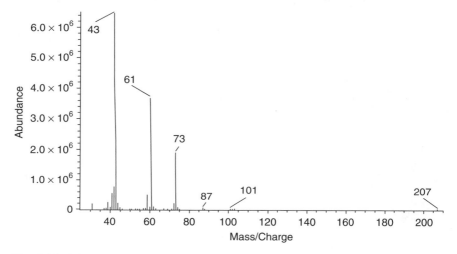

Fig. 4.14 Mass spectrum from the chromatogram of the non-odorous bag in the region of 7.2 min.

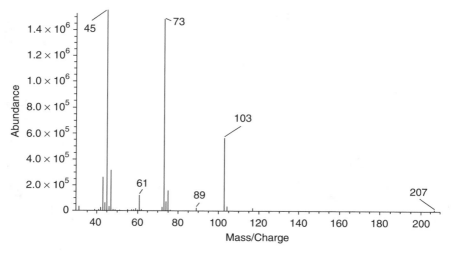

Fig. 4.15 Mass spectrum from the chromatogram of the odorous bag in the region of 7.2 min.

presence of significantly more of the 1,1-diethoxyethane in the odorous bag compared to the non-odorous bag is clearly shown in Figs 4.16 and 4.17 where the 103 m/z ion has been extracted.

Finally, to prove that the chemical responsible for the odour had been correctly identified, a sample of 1,1-diethoxyethane was injected and the retention time, mass spectrum and odour from the sniffer port compared

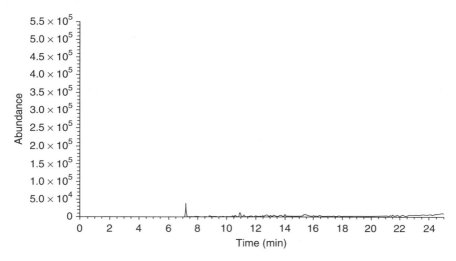

Fig. 4.16 Chromatogram from non-odorous bag with 103 m/z extracted.

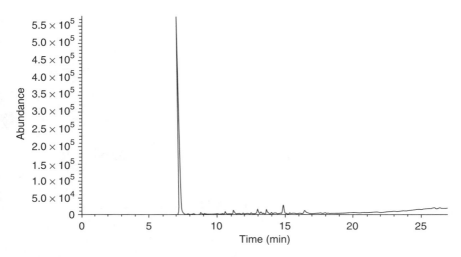

Fig. 4.17 Chromatogram from odorous bag with 103 m/z extracted.

96   Taints and off-flavours in food

with that obtained from the odorous bags (Figs 4.18 and 4.19). This confirmed that the cause of the odour was 1,1-diethoxyethane.

The manufacturer of the bags suspected that an unwanted chemical reaction was the cause of the generation of the 1,1-diethoxyethane. A titanium-based chelate was known to have been used as an adhesion promoter and it was suspected that this may have catalysed formation of the 1,1-diethoxyethane. Analysis of the bags by purge and trap had identified the

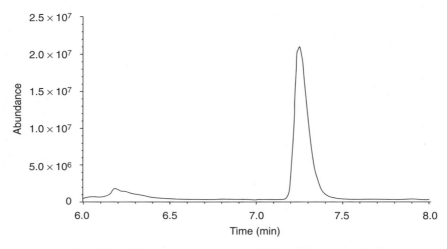

Fig. 4.18   Chromatogram from 1,1-diethoxyethane standard.

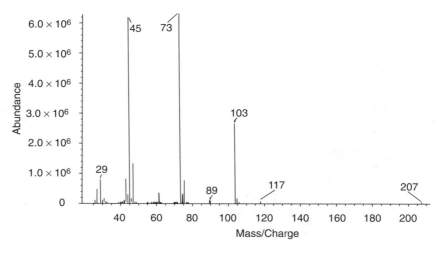

Fig. 4.19   Mass spectrum obtained from 1,1-diethoxyethane standard.

presence of significant amounts of acetaldehyde and ethanol. In the presence of a suitable catalyst, 1,1-diethoxyethane is known to be generated from these starting materials (March, 1977; Adkins and Nissen, 1922). Experiments demonstrated that 1,1-diethoxyethane is indeed generated by heating the adhesion promoter with ethanol and acetaldehyde at the bag processing temperatures. However, subsequent analysis of the solvents used to dilute the inks on the press revealed the presence of relatively large amounts (a few hundred parts per million) of the 1,1-diethoxyethane. The source of the 1,1-diethoxyethane was therefore more likely to be from the use of impure solvents. This illustrates that although a mechanism of odour generation may have been proved in the laboratory, it is very dangerous to assume that this is the cause. It is also important to reflect on how likely the source of the odour is. Titanium adhesion promoters are widely used and 1,1-diethoxyethane is not a common cause of odour on this type of packaging. Impure raw materials are often the root cause of taint problems.

4.9.3 Case study 3: packaged cheese

A final example illustrates the importance of a thorough, self critical, objective approach and the need to analyse packaging components before drawing firm conclusions from laboratory work. A complaint of a chemical odour and taste was reported in packed cheese. The pack construction consisted of a PVC/PE tray and a PE/PET lid. The lid had at least two layers of PE bonded using an adhesive. The lid was heat sealed to the filled tray during cheese packing. Upon opening, the lid was designed to tear at the PE/PE interface exposing the lid adhesive, thus providing a re-sealing lid. A pungent, chemical pine odour was reported upon opening the cheese packs. Dynamic headspace analysis on the complaint cheese pack (Fig. 4.20) identified the presence of 1,3-pentadiene, α-methyl styrene, limonene and a series of related terpene compounds.

The 1,3-pentadiene was suspected to arise from biological decarboxylation of sorbic acid preservative used in the cheese (Saxby, 1996). The sorbic acid was detected in the headspace from the cheese packs. Dynamic headspace analysis of an unused pack (Fig. 4.21) demonstrated the absence of the 1,3-pentadiene and the presence of the other suspected tainting compounds.

The complaint odour was different from *trans*-1,3-pentadiene and this was eliminated as the cause. Odour panel tests were carried out on the filled and unfilled packs. The complaint odour could be smelt on the tray from empty packs, although there was an additional stale beer odour. This stale beer odour was found to be the main tray odour after standing for 10 min at room temperature and was not an intense odour. The complaint odour was not present on the inside of the lid. Dynamic headspace analysis on the unused tray (Fig. 4.22) identified the same suspected tainting compounds identified above.

98   Taints and off-flavours in food

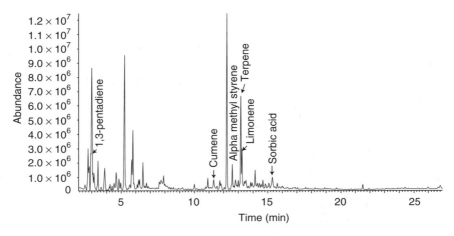

Fig. 4.20   Dynamic headspace GC-MS on an odorous pack.

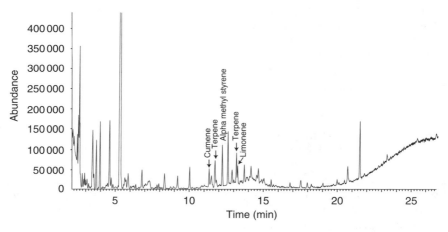

Fig. 4.21   Dynamic headspace GC-MS on an unused problem pack.

The complaint odour could not be smelt at the odour port on any of the samples. A series of individual solvent-like odours could be smelt but not the overall complaint odour. Samples of unconverted base and top web from which the tray and lid were made were not initially available for analysis. In their absence, the packaging components from the packs were investigated further. The complaint odour and the additional stale beer odour could be

Packaging materials as a source of taints 99

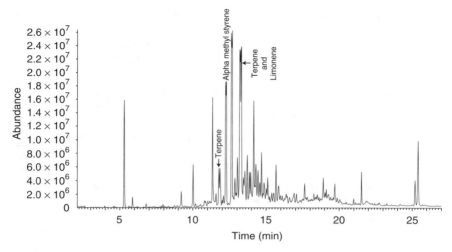

Fig. 4.22 Dynamic headspace GC-MS on the unused tray.

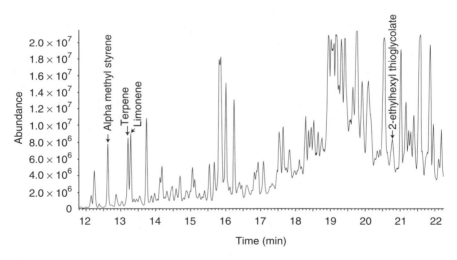

Fig. 4.23 Dynamic headspace GC-MS of a pentane-extracted residue from an unused tray.

extracted with pentane. On-column injection of the extract resulted in the strong stale beer odour being smelt at the odour port at 20.7 min. Dynamic headspace analysis of a pentane extract is shown in Fig. 4.23.

The compound 2-ethylhexyl thioglycolate was identified, eluting at 20.7 min. The 2-ethylhexyl thioglycolate was suspected to be derived from a tin stabiliser because of the presence of 2-ethylhexanol and previous experience in the laboratory of odours generated from PVC. The tray was subjected to a series of chemical analyses and the tin stabiliser identified as an alkyl tin stabiliser of the type $(R)_n$-Sn-$(SCH_2CO_2C_8H_{17})_n$, where $n = 1, 2$ or 3 and R = a methyl or octyl group. A small quantity of commercial octyl tin stabiliser was heated with 0.1 N hydrochloric acid in dioxane and 2-ethylhexyl thioglycolate was generated. 2-Ethylhexyl thioglycolate was therefore confirmed to arise from decomposition of the tin stabiliser by the action of hydrochloric acid from the PVC and heat during thermoforming of the PVC tray.

Odour panel tests with 2-ethylhexyl thioglycolate and the trays confirmed that although responsible for a slight background tray odour it was not the complaint odour smelt when the trays were first opened. Attention turned to tracing the source of the α-methyl styrene, limonene and isoprene compounds identified in the odorous packs. Static headspace analysis of the pack components obtained by delaminating an empty pack established that the adhesive used in the lid laminate was the source. When the unconverted tray and lid material was received for analysis, it was immediately obvious from the odour of the lid material that the lid was the source of the odour. Dynamic headspace analysis on the lid material gave a series of odours at the odour port that were similar but not identical to the complaint odour. The reason for this was that the complaint odour was a combination of odours and was not due to one individual chemical. The compounds giving these odours were indeed the α-methyl styrene, limonene and related terpenes identified in the complaint packs. The overall odour of the unconverted lid material was identical to the complaint odour. The cause of the odour was therefore due to the presence of α-methyl styrene, limonene and the series of related terpenes in the adhesive used in the lid. These compounds had migrated from the lid into the empty tray. The slight odour from the thermoformed trays could be explained by the presence of the alkyl tin stabiliser and 2-ethylhexyl thioglycolate.

The fact that the tainting compounds were present at higher concentration on the tray rather than the lid could have resulted in the erroneous conclusion that they originated from the tray. Although providing some additional information, tin stabiliser analysis and the identification of 2-ethylyhexyl thioglycolate was a 'red herring' and a distraction from the main cause of the odour.

This example shows that the typical PVC odour in food trays, which is quite common and usually perfectly acceptable, may often be explained by the presence of octyl tin stabilisers and also thioglycolate esters that are produced from the tin stabilisers during thermoforming of the PVC sheet into trays. As a rule odour problems are usually traceable to lidding laminates rather than PVC/PE trays. The following are useful points to remember:

Packaging materials as a source of taints 101

1 Check that conclusions have been proved by experimental fact. It is possible that an incorrect assumption has been made.
2 Establish the true construction of the packaging.
3 Always obtain a complete set of samples to work on, including the separate packaging components or unconverted films.
4 When the tainting chemical(s) have been identified, verify that they are the cause using a panel test. Compare concentrations found with literature or laboratory-determined taint thresholds.

Recently a catty urine odour in printed PVC film was reported to Pira. It was not possible to investigate this case fully. However, it was suspected that the cause of the odour was 4-methy-4-mercaptopentan-2-one and that this compound may have been generated by the reaction of mesityl oxide with a component of the film, possibly associated with the alkyl tin stabiliser used in the PVC. The evidence for this was that the same catty urine odour could be generated by applying mesityl oxide to the surface of the unprinted film and storing it in a glass bottle containing a drop of concentrated hydrochloric acid for 2 days at 40°C. Mesityl oxide and diacetone alcohol were present on the printed odorous film. The tin stabiliser used was an alkyl tin thioester. The diacetone alcohol was a component of the ink. The film had not been used and so had not been in contact with meat or other obvious sulfur-containing sources. The case highlights the importance of excluding even traces of diacetone alcohol or acetone in food packaging. It is worth also noting that analysis for this compound can be difficult because of the very low concentrations required to give the odour. Difficulties in the analysis are reported by Franz *et al.* (1990). Invariably selected ion monitoring (m/z 132) is required. Laboratory preparation of the compound by the action of hydrogen sulfide upon mesityl oxide can produce surprisingly low yields. The compound can be purchased as a 1% solution in polyethylene glycol from Oxford Chemicals Ltd.

4.10 Preventing taints

4.10.1 The legislation
The first European Council Directive laying down the basic rules for food contact materials and articles was adopted in 1976. This became known as the Framework Directive, 76/893/EEC, with the title 'On the approximation of the laws of the member states relating to materials and articles intended to come into contact with foodstuffs'. Article 2 Directive 76/893/EEC specifies requirements for food contact materials as:

> Materials and articles shall be manufactured in accordance with good manufacturing practice, so that, under their normal and foreseeable

conditions of use, they do not transfer their constituents to foods in quantities, which could:
- endanger human health
- bring about a deterioration in the organoleptic characteristics of such food or an unacceptable change in its nature, substance or quality.

In the UK, these basic requirements for food contact materials and articles from Directive 76/893/EEC were incorporated into UK law by Statutory Instrument (SI) No 1523 (1987); The Materials and Articles in Contact with Food Regulations. A new EC Framework Directive 89/109/EEC was introduced in 1988 but the above health and organoleptic requirements were retained.

In the United States, similar regulations to those in Europe exist to prohibit packaging affecting the organoleptic qualities of packaged food. The Federal Food, Drug and Cosmetics Act, Section 402(a)(3) specifies that any 'indirect food additive' originating from packaging should not transfer to food 'odour or off-taste rendering it unfit for consumption'.

4.10.2 Test methods

A number of sensory-type tests are described by Tice (1996). A brief summary is given here of the most useful tests for taint prevention. The purpose of the following tests is to ensure that a taint does not occur in the packaged food as a result of chemicals migrating in sufficient quantities from the packaging.

Panel tests can be odour and taste tests. Ideally they are best done in a neutral odour test room (DIN 10962). The odour tests include the German DIN Standard 10955, 1993, BSI Standard BS EN 1230-1:2001 and the American ASTM standards E462-84 (reapproved 1989) and E619-84 (reapproved 1989).

The EN standard 1230-1:2001 for odour assessment is carried out in the following manner. Test portions of area $6\,dm^2$ are placed in a wide neck, flat bottom glass container. The containers are sealed and maintained at 23°C for 20 to 24 hours. At the end of the exposure time the container is opened and the odour intensity assessed using the following scale:

0 = no perceptible odour
1 = odour just perceptible (but still difficult to define)
2 = moderate odour
3 = moderately strong odour
4 = strong off-odour.

Half-number scores may be recorded if considered necessary by a panellist. The median is calculated from all the individual scores. One individual result may be discarded if it differs from the median by 1.5 or more.

If there are fewer than six consistent results the test is repeated with fresh samples.

DIN standards have been used for odour testing. These are either a ranking in order of odour intensity (DIN 10963), or a paired comparison test against a reference (DIN 10954).

For taste testing on packaging, there is the standard EN 1230-2:2001 part 2. Companies often use their own test procedures, which are usually based on the Robinson test. The Robinson test involves exposure of chocolate to the packaging followed by a sensory panel test. In the EN 1230-2:2001 part 2 test the packaging material ($6\,dm^2$) is placed in a sealed container separated from the test food. The humidity of the air inside the container is maintained at between 53 and 75% using selected saturated salt solutions. After 44 to 48 hours at 20°C the test food is removed. A sensory test is then used to determine whether a taint has transferred to the chocolate.

Three types of panel tests may be used in EN 1230-2:2001 part 2 procedure. These are the triangle test, the extended triangle test and the multi comparison test. In the extended triangle test, one portion of the test chocolate that has been exposed to the packaging under test, is placed next to two portions of control food exposed to similar conditions without the presence of the packaging material. A panel of 10 assessors is used. The assessors are asked to pick out the test food and rate it on the following scoring system:

0   no perceptible off-flavour
1   off-flavour just perceptible (still difficult to define)
2   moderate off-flavour
3   moderately strong off-flavour
4   strong off-flavour.

The median score is calculated from all the individual scores discarding any scores given for the control test chocolate. If more than five assessors identify the test food it is concluded that at the 95% confidence level a taint has been transferred. Results are not easily compared on a numerical or statistical basis with results from other laboratories especially if they operate different scoring systems. As a quality control tool the test can be very useful. It is important to maintain a reliable set of panellists. The triangle test is carried out in the same way as the extended triangular test except that no intensity ratings are required. In the multi comparison test the panellist is required to score the intensity against a known test portion used as a control which is assigned a taint intensity of 0.

The triangle test is used when a taint intensity rating is not required. The extended triangular test is useful for the comparison of a number of packaging samples. The multi comparison test is useful for large numbers of samples.

Residual solvent analysis is widely carried out. The best technique is static headspace gas chromatography as described in Section 4.7.2. The procedure to be followed is set out in a current European Committee for Standardisation (CEN) draft standard Draft prEN 13628-1 (July 1999) and Draft prEN 13628-2 (July 1999). Part 1 is the quality control monitoring procedure. Part 2 is the more accurate definitive test procedure. There are no limits set for retained solvents in packaging. However, there is an industry guide limit of individual solvents not exceeding $5\,\mathrm{mg\,m^{-2}}$ and total solvent residues not exceeding $20\,\mathrm{mg\,m^{-2}}$. The packaging should be selected in the knowledge of any promotional items required. It is useful to have a specification for the packaging. Items to include in this specification are listed in Section 4.10.3 below. Simple taint panel tests can be carried out to ensure that the packaging cannot transfer taint to the food. Accelerated storage tests are also useful to assess the likelihood of a taint problem occurring.

4.10.3 Practical steps to avoid the occurrence of taints

Many taint problems could be avoided by setting and ensuring compliance with suitable packaging specifications. Specifications could include the following:

- Residual solvent: $<20\,\mathrm{mg\,m^{-2}}$
- Individual solvents (esters): $<5\,\mathrm{mg\,m^{-2}}$
- Promotional items in food packaging: flow wrapped with PVDC coated OPP (oriented polypropylene) unless the absence of a taint transfer has been proved
- Gauge of promotional flow wrap: resistant to puncture by promotional item
- Packaging and promotional items: passes panel taint tests
- Inks: low odour from a reputable supplier
- Solvents: high purity from a reputable supplier
- Paper: preferably not recycled for direct food contact
- Plastic packaging: certified for food use, tested to ensure compliance with US or European law for the conditions of use.

Any new packaging or change in packaging format should be checked with a trial production run and a panel test.

4.11 Developments in taint monitoring: electronic noses

In the mid-1990s, electronic noses attracted interest in the packaging industry for their potential to control taints arising from packaging. Developments in computer signal processing coupled with the development of new sensors offered the possibility of building sensor arrays that could be trained to recognise the 'pattern' or 'signature' of a taste or smell. The distinction between existing analytical equipment is that instead of physically separating a chemical causing a taint from all the others present using a gas or liquid chromatograph, the overall signal from a tainted sample could be recognised as being different to a taint-free sample. In the same way a human brain is able to decide that a taint or odour is different, without any separation of the bad taste or odour causing compounds being required.

Pattern recognition algorithms are used to train a computer to categorise the signals it receives from good (non-odorous) and bad (odorous) samples. Sensors based on conducting polymers and/or metal oxides are used to generate the signal. Both types involve a change in the electrical resistance of the sensors. The metal oxides involve oxidative changes of the tainting compounds at the sensor surface. The conducting polymer involves binding the tainting compound on the sensor surface which has an effect on electrical conductivity. The metal oxide is generally capable of providing a much larger change in response than conducting polymers. Acoustic wave sensors have been developed where an electric oscillating standing wave is set up in the sensor. Compounds are absorbed onto the sensor surface, resulting in a change in mass and therefore the oscillator frequency. All the sensor types suffer disadvantages: the conducting polymers respond to water; and metal oxides and conducting polymers are poisoned by some compounds. A review of conducting polymers was carried out by Adeloju and Wallace (1996). Ide *et al.* (1997) describes a quartz microbalance electronic nose-type sensor.

The availability of very cheap quadrapole mass spectrometers has renewed interest in this technology being able to recognise taint patterns. The approach taken is to use a six-port sample valve to introduce a small volume of air from an equilibrated headspace above the sample directly into the mass spectrometer whilst it is scanning. After a few seconds the mass spectrometer response stabilises and the whole scan of mass fragments is stored in a statistical software package. The six-port valve is then switched back to a constant stream of helium. The mass range scanned is quite narrow, for example between 30 and 200 m/z, the reason being that base peaks contribute the bulk of the signal, and the considerably less abundant but more characteristic molecular ions contribute only a tiny proportion of the signal. The significance of a weak molecular ion is lost in the comparatively high background. The pattern recognition software is used to pick out fragmentation differences between good and bad samples. The same approach can be used for a liquid sample injected into a liquid stream to investigate tastes. In this case the instrument functions as an electronic tongue. Such equipment is more costly because of the interface problems involved in removing water from the sample before it enters the mass spectrometer.

At present, electronic nose technology is not at the stage where it can be generally applied to quality control tests and it offers no advantages over GC-MS for investigative work. The cost of the equipment is still much too high for it to become widely adopted in packaging production. The current state-of-the-art is summarised by Gibson *et al.* (2000), whose article lists the main sensor types and suppliers.

4.12 Tracing the cause of a packaging taint

As much information as possible about the problem, history of its occurrence and nature should be obtained. Is it a taste or an odour? What is the odour or taste described as? What is the production process? If samples are sent by post they should be wrapped in aluminium cooking foil as separate blocks of sample to prevent cross contamination and loss of volatiles. Much can be learnt from simple panel tests using assessors selected for their proven above average ability to detect taints. Tainted and untainted samples can be identified in this way before beginning the more costly analytical work. The following steps should be taken as soon as a taint has been reported:

1. Act quickly to prevent further distribution of the product if possible.
2. Set up a team representing the key functions involved, e.g. the printing supervisor, technical manager.
3. Engage a consultant with the capability to advise and carry out chemical analysis and or taint panel tests at their own laboratories.
4. Obtain complaint samples, previous good samples and any raw materials still existing for these two sample sets.
5. Confirm the nature of the problem, i.e. odour or taste description.
6. Establish and maintain a case history.

It is important that the consultant has the analytical equipment and experience to carry out the investigation in his or her laboratory. Taint investigation is not suited to a paper-based exercise where specific analysis can be contracted out to laboratories, and the raw data interpreted. It is most unlikely that such an approach will deliver the correct solution. Success often relies upon selection of the most appropriate sample preparation technique for the particular sample and observation of important clues by the analyst.

When a particular chemical has been identified in the tainted sample at higher concentration than in the untainted sample, the concentration should be compared to a taint threshold. It is important to choose a threshold value in the literature that has been measured in a similar food. Tice (1996) describes how the taint threshold for styrene varies widely depending upon the food. For water the threshold is 0.022 ppm, for whole milk it is 1 ppm and for butter it is 5 ppm. The odour threshold in air is 0.05 ppm. Taint threshold values for styrene and other tainting compounds in a range of foods are listed by Ackermann *et al.* (1995). Simply finding a potential tainting compound in the packaging is not sufficient to conclude that it is the cause. For example, concentrations of free styrene can be 200 ppm in

Packaging materials as a source of taints 107

polystyrene polymers without significant migration into many foods occurring. Once the chemical(s) causing the taint have been identified, a great deal of work is then often required to establish where they originated, how they came to be present, and what action is required to prevent a reoccurrence.

4.13 Future packaging trends affecting taints

New technology is continually being applied to food packaging. Some developments that are likely to emerge in the next few years are discussed below.

4.13.1 Cationic inks

Although lower odour compared to UV radical curing is sometimes claimed for cationic inks, lower odour is not the principal reason that will drive this technology forward. A mature technology, UV radical curing is quite capable of achieving a low odour product. At present cationic technology accounts for no more than a few per cent of the radiation curing technology used in packaging. Cationic-cured inks offer the advantage of high gloss and less shrinkage than conventional free radical UV-cured inks with better adhesion on metal surfaces. Cationic inks can provide high grease resistance on packaging such as pet food bags and garlic bread bags.

The UV light is used to initiate the cationic curing process. Once the cure has been initiated, curing continues in the dark. The photoinitiator does not itself take part in the curing process. The photoinitiator is used to produce strong acids such as hydrofluoric acid, which cause the actual curing. The development of the technology has been held up by the lack of commercially available photoinitiators at competitive prices. The most widely used photoinitiators for food packaging applications have been triaryl sulfonium salts. Diphenyl sulfide (highly odorous) can be produced as a by-product in the polymerisation process of some cationic inks. Following concerns over the possibility of benzene being produced during curing, non-benzene-producing initiators have been introduced. An excellent review of cationic inks is provided by Crivello and Dietliker (1998).

4.13.2 Digital printing

Computer technology has now developed to the point where artwork can be designed on a PC and sent straight to the printer, without the need for a plate or roller to transfer the image. The inks are transferred to the paper or plastic using processes somewhat similar to a photocopier. The inks are usually powders that are fused by a heater onto the surface. The use of this technology is likely to expand in the future particularly for short print runs and items such as labels. The reasons for this are the speed with which an image can be designed, and reviewed and printed, or transmitted. The image can be sent anywhere in the world to be printed. Storage and reuse of the images is much easier. Considerable development and investment in the technology, however, is required for it to compete with the line speed and costs achieved with conventional flexo or gravure printing. The advantage as far as taint risk is concerned is that the printing process is solvent free.

4.13.3 In-store printing
The advantage of in-store printing technology is the flexibility it provides, for example to enable a store to add promotional designs onto small batches of product. The technology could potentially involve additional risk as far as taint is concerned. The printing may not be carried out under tightly monitored and controlled conditions.

4.13.4 Co-extrusions
The continued shift towards co-extrusion structures and away from laminates manufactured using solvent-based adhesives will reduce the risk of taints. Layered structures are produced by extruding a whole series of layers on top of each in a molten form. As it cools the layers fuse, forming a single structure without the use of any adhesive.

4.13.5 Smart and active packaging
Smart and active packaging technology has more to do with the prevention of off-odour and taste as a result of food spoilage than taint occurring as a result of a packaging interaction. However, taint scavengers are being developed for incorporation into packaging to trap potentially tainting compounds from the packaging components within the packaging. These scavenging compounds are often crystalline powders such as xeolites, which have a cage-like molecular structure that can trap a range of tainting compounds. An example of an application is 'bag in a box'-type packaging for drinks. The problem with these taint scavengers is their lack of specificity. Taste scalping can therefore be a problem, with flavour compounds being removed from the food. The result can be an imbalance of flavours which manifests itself as a taint.

4.14 Sources of further information and advice
An essential tool for anyone working with taint is a compilation of the most comprehensive taint and odour threshold values published. Two very useful ones are Fazzalari (1978) and Saxby (1992).

There is a lot of information available now on the Internet. The status of electronic nose technology can be followed on the EU sponsored web site http://nose.uia.ac.be. The web site www.leffingwell.com lists a range of chemicals with their odour and taste thresholds. The site http://chemfinder.cambridgesoft.com has a useful database of commercially available chemicals. This may be of use in obtaining a chemical as an analytical calibration standard. The site www.ChemWeb.com has extensive databases on chemicals and abstracts of published articles in the main scientific journals. Abstracts and articles on scientific publications are also available from the site www.info.sciencedirect.com. This site has a search engine designed for scientific information retrieval. The web search engines such as www.google.co.uk can be used to find up to date information on newly added sites. A list of keywords is entered, for example 'odour', 'packaging taints', relevant sites are then listed and these sites can be opened by merely selecting them from the search list.

4.15 References

ACREE T E, BUTTS R M, NELSON R R and LEE C Y (1976), 'Sniffer to determine the odour of gas chromatographis effluents', *Anal Chem*, **48** (12), 1821.
ACREE T E, BARNARD J and CUNNINGHAM D G (1984), 'A procedure for the sensory analysis of gas chromatographic effluents', *Food Chem*, **14**, 273–86.
ACREE T E (1997), 'GC/olfactometry GC with a sense of smell', *Anal Chem* News Features, 171–5.
ACKERMANN P M (1995), *Foods and Packaging Materials–Chemical Interactions*, in Ackermann PM, Jagerstad M and Ohlsson T (eds), The Royal Society of Chemistry, Special Publication No 162.
ADELOJU S B and WALLACE G G (1996), 'Conducting polymers and the bio-analytical sciences: new tools for bimolecular communications, A review', *Analyst*, **121**, 699–703.
ADKINS H and NISSEN, B H (1922), 'A study of catalysis in the preparation of acetal', *J Am Chem Soc*, **44**, 2749–55.
ASTM (1989a), Standard: E 462-84 (re-approved 1989), 'Standard Test Method for Odour and Taste Transfer from Packaging Film'.
ASTM (1989b), Standard: E 619-84 (re-approved 1989), 'Standard Practice for Evaluating Foreign Odours in Paper Packaging'.
BADINGS H T and DE JONG C (1985), 'Automatic system for rapid analysis of volatile compounds by purge-and-trapping/capillary gas chromatography', *J High Resolu Chromatogr Commun*, **8**, (November), 755–63.
BOUSETA A and COLLIN S (1995), Optimised Likens–Nickerson methodology for quantifying honey flavours, *J Agric Food Chem*, **43**, 1890–7.
BROWN, R H (1996), 'What is the best sorbent for pumped sampling–thermal desorption of volatile organic compounds? Experience with the EC sorbents project', *Analyst*, **121**, 1171–5.
BRAVO A, HOTCHKISS H and ACREE T (1992), 'Identification of odour-active compounds resulting from thermal oxidation of polyethylene', *J Food Chem*, **40**, 1881–5.
BRAVO A and HOTCHKISS J H (1993), 'Identification of volatile compounds resulting from the thermal oxidation of polyethylene', *J Appl Polym Sci* **47**, 1741–8.

BUCHALLA R, SCHUTTLER C and BOGL K (1993), Effects of ionising radiation on plastic food packaging: A review, *J Food Protection*, **56** (11), 998–1005.
BUCHALLA R, BOESS C and BOGL K W, (1998), Radiolysis products in gamma-irradiated plastics by thermal desorption GC-MS, parts 1 and 2, *Bundesinstiut fur gesundheitlichen Verbraucherschutz und Veterinarmedizin*, **4**, 32–7.
CLARK T J and BUNCH J E (1997), Derivatization solid phase micro-extraction gas chromatographic-mass spectrometric determination of organic acids in tobacco, *J Chromatogr Sci*, **35**, 209–12.
CRIVELLO J V and DIETLIKER K (1998), *Photoinitiators for Free Radical Cationic and Anionic Photopolymerisation*, 2nd edn, Bradley G (ed), Vol. 3, John Wiley and Sons, London.
DIETZ F and TRAUD J (1978), Odour and taste threshold concentrations of phenolic compounds, *Gas-Wasserfach, Wasser–Abwasser*, **119** (6), 318–25.
DUEK-JUN AN and HALEK G W (1995), Partitioning of printing solvents on chocolate, *J Food Sci* **60** (1), 125–7.
EN 1230-1:2001 Paper and Board Intended for Contact With Foodstuffs-Sensory analysis-part 1:Odour, October 2001.
EN 1230-1:2001 Paper and Board Intended for Contact With Foodstuffs-Sensory analysis-part 2:Off-flavour (taint), October 2001.
FAZZALARI F A (1978), '*Compilation of Odour and Taste Threshold values data*', American Society For Testing and Materials ASTM Data Series DS 48A, Philadelphia.
FRANZ R, KLUGE S, LINDER A and PIRINGER O (1990), 'Cause of catty odour formation in packaged food', *Packaging Technol Sci*, **3**, 89–95.
GIBSON T, PROSSER O and HULBERT J (2000), 'Electronic noses: an inspired idea?', *Chem Ind*, **17** (Aprl), 287–9.
HUBER M, RUIZ J and CHASTELLAIN F (2002), Off-flavour release from packaging materials and its prevention: a foods company's approach, *Food Additives and Contam*, 19, (Supplement), 221–8.
HOFF A and JACOBSSON S (1981), Thermo-oxidative degradation of low density polyethylene close to industrial processing conditions, *J Appl Sci*, **26**, 3409–23.
IDE J, NAKAMOTO T and MORIIZUMI T (1997), An odour-sensing system for use in measuring volatiles in flavour and fragrances using OCM, in Swift KAD (ed), *Flavours and Fragrances*, Royal Society of Chemistry, Cambridge, pp 87–95.
ISO (1992), Glossary of Terms Relating to Sensory Analysis, ISO standard 5492.
KILCAST D (1990), Irradiation of packaged food, in Johston D E and Stevenson M H (eds), *Food Irradiation and the Chemist*, The Royal Society of Chemistry, Cambridge, UK Special Publication No 86, pp 140–152.
KILCAST D (1996), 'Sensory evaluation of taints and off-flavours', *Food Taints and Off-flavours*, 2nd edn, Saxby MJ (ed), Blackie Academic and Professional, London, pp 1–40.
KOSZINOWSKI and PIRINGER O (1986), 'Evaluation of off-odours in food packaging–the role of conjugated unsaturated carbonyl compounds', *J Plastic Film Sheeting*, **2**, 40–50.
KURSCHNER K (1926), Zur Chemie der lignin-korper, Sammlung Chem.u.ChemTech. Vortrage, **28** (71), Stuttgart.
LIKENS S T and NICKERSON G B (1964), Detection of certain hop oil constituents in brewing products, *Proc Am Soc Brew Chem* **5**, 5–14.
MARCH J (1977), *Advanced Organic Chemistry, Reactions, Mechanisms, and Structure*, 2nd edn, McGraw-Hill, Tokyo, pp 810–12.
MARIN A B, ACRE T E, HOTCHKISS H and NAGY S (1992), Gas chromatography–olfactometry of orange juice to assess the effects of plastic polymers on aroma character, *J Agric Food Chem* **40**, 650–4.
MCGORRIN R J, POFAHL T R and CROASMUN W R (1987), Identification of the musty component from an off-odour packaging film, *Anal Chem*, **59**, (18), 1109–12.

PERKIN ELMER (1993), Thermal Desorption Data Sheet Number 10: *A Guide to Adsorbent Selection*, The Perkin Elmer Corporation, Post Office Lane, Beaconsfield, Bucks HP9 1QA.
PRETI G, GITTELMAN T S, STAUDTE P B and LUITWEILER P (1993), *Anal Chem*, **65**, (15), 699–702.
REINECCIUS G (1991), Off-flavours in foods, *Food Sci Nutrition*, **29** (6), 381–402.
ROFFEY C G (1982), *Photo Polymerisation of Surface Coatings*, John Wiley and Sons, Chichester.
SANDERS R A (1983), Identification of cyclic acetals in polyols by mass spectrometry, *Anal Chem*, **55**, 1194–7.
SAXBY M J (1992), '*Index of Chemical Taints*', updated edition December 1992, Leatherhead Food Research Association.
SAXBY M J (1996), A survey of chemicals causing taints and off-flavours in food, in Saxby M J (ed), *Food Taints and Off-flavours*, 2nd edn, Chapman and Hall, Glasgow.
SUPLECO (1994), *Fast Analysis of Volatile Organic Compounds by Solid Phase MicroExtraction/Capillary GC*, Application Note 56.
SUPLECO (1995), *Solid Phase Microextraction: Solventless sample preparation for monitoring flavour and fragrance compounds by capillary gas chromatography*, Bulletin 869.
SUPLECO (1996), *Polyacrylate Film Fibre for Solid Phase Micro-extraction of Polar Semivolatiles from Water*, Application Note 17.
SUPLECO (1997), *Optimizing SPME: Parameter to control to ensure consistent results*, Application Note 95.
TICE P A and OFFEN C P (1994), Odours and taints from paperboard food packaging, *TAPPI J*, **77**, (12), 149–154.
TICE P (1996), Packaging materials as a source of taints, in Saxby M J (ed), *Food Taints and Off-flavours*, 2nd edn, Chapman and Hall, Glasgow.
VILLBERG K, VEIJANEN A, GUSTAFSSON I and WICKSTROM K (1997), Analysis odour and taste problems in high-density polyethyelene, *J Chromatogr A*, **791**, 213–19.
WHITFIELD F B, HILL J L and SHAW K J (1997), 2,4,6-tribromoanisole: a potential cause of mustiness in packaged food, *J Agric Food Chem*, **45**, 889–93.
WHITFIELD F B, LY NYUYEN T H and TINDALE R C (1989), Shipping container floors as sources of chlorophenol contamination in non-hermetically sealed foods, *Chem Ind*, 458–9.
WERKHOFF P and BRETSCHNEIDER W (1987), Dynamic headspace gas chromatography: Concentration of volatile components after thermal desorption by intermediate cryofocusing in a cold trap, *J Chromatogr*, **405**, 87–98.

5
Microbiologically derived off-flavours

F. B. Whitfield, Food Science Australia, Australia

5.1 Introduction

Microbiological spoilage of food is well known to all consumers and one of the most characteristic indicators of this form of spoilage is the development of an unpleasant odour or flavour in the product. Such microbial contamination can occur as a result of poor handling and storage of the food by the consumer. However, in major outbreaks the contamination has usually taken place before distribution. Where such contamination occurs it can involve only a relatively small proportion of a processing run, or at worst affect one or more day's production. As the initial microbiological contamination could have been slight, the problem often only becomes apparent some days or weeks after production. It is these incidences that usually involve investigation by microbiologists and chemists. Before the 1980s, these two disciplines tended to work in isolation and rarely was there a unified approach to solve such problems. Consequently, the link between the contaminating organisms and the compounds responsible for the off-flavour was seldom achieved. Fortunately, since the early 1980s a more co-ordinated approach has developed. As a result, a greater number of organisms and their metabolites are now being identified as the cause of characteristic off-flavours in foods.

In previous reviews of this topic[1,2] the trend has been to discuss the causes of microbiologically derived off-flavours according to the food in which they occurred. However, in this chapter a different approach will be taken and problems will be discussed according to the organism responsible for the off-flavour. The organisms usually associated with such problems are bacteria, fungi and to a lesser extent, yeasts. In discussing off-flavours

Table 5.1 Microorganisms discussed in this chapter

Aerobic bacteria	*Acinetobacter*
	Alcaligenes (Achromobacter)
	Altermonas
	Alicyclobacillus
	Moraxella
	Pseudomonas
Facultative anaerobic bacteria	*Bacillus*
	Brochothrix
	Serratia
	Rahnella
	Yersinia
Anaerobic bacteria	*Clostridium*
Actinomycetes	*Streptomyces griseus*
Fungi	*Aspergillus*
	Eurotium
	Penicillium

of bacterial origin, the organisms will be divided into aerobic, facultatively anaerobic and anaerobic bacteria. Actinomycetes will be discussed in a separate section but cyanobacteria will not be covered in this review as the literature is far too extensive. Fungi will be discussed according to genus (Table 5.1).

The foods covered by this review will include raw and processed meats, raw seafoods, dairy products, fruit and vegetable products, and cereal and cereal products. Some key foods and common off-flavours associated with them are shown in Table 5.2. The table shows the microorganisms and compounds responsible for these off-flavours. This selection of foods covers a wide range of water activities (a_w), pH, storage temperatures and atmospheric conditions. However, food spoilage microorganisms are extremely versatile and can grow under most conditions associated with the production, handling and storage of fresh and processed foods. As a consequence, most foods at some stage before consumption are susceptible to microbial contamination and possible off-flavour development. Accordingly, this chapter will not only describe the microorganism and metabolites responsible for an off-flavour, but also the conditions that favour production of the offending compound.

5.2 Bacteria

Bacteria capable of food spoilage can grow at a_w values from 0.75 to 1, pH values from 2.5 to 10, temperatures from −2 to 75°C and under a variety

Table 5.2 Microbiologically derived off-flavours in particular foods

Food	Microorganisms	Compounds produced	Off-flavour
Milk and cheese	*Streptococcus* spp. *Pediococcus* spp. *Leuconostoc* spp. *Lactobacillus* spp. *Pseudomonas* spp. *Yersinia intermedia*	Lactic acid, butanoic acid, hexanoic acid, octanoic acid, decanoic acid	Acid
	Streptococcus lactis biovar *maltigenes*	3-Methylbutanal, 2-methylpropanol	Malty/cooked/caramel
	Pseudomonas fragi, P. putida, Yersinia intermedia	Ethyl butanoate, ethyl hexanoate, ethyl octanoate	Fruity
	P. fluorescens, P. fragi, P. putrefaciens	Peptides	Bitter
	P. taetrolens	2-Methoxy-3-isopropyl pyrazine	Musty/potato
Beef, pork and lamb	*Brochothrix thermosphacta* *Lactobacillus* spp.	Diacetyl (2,3-butanedione), acetoin, 3-methylbutanol, 2-methylpropanol	Buttery/cheesy
	Pseudomonas spp. In particular *P. fragi. Moraxella* spp. *B. thermosphacta* and Enterobacteriaceae/*Pseudomonas* spp.	Ethyl acetate, ethyl butanoate ethyl hexanoate, ethyl octanoate	Sweet/fruity
		Ethyl 2-methylbutanoate ethyl 3-methylbutanoate 2-methylbutyl acetate	Fruity
	Pseudomonas perolens *Pseudomonas taetrolens*	2-Methoxy-3-isopropylpyrazine	Musty potato-like
	Enterobacteriaceae *Shewanella putrefaciens*	Methanethiol, dimethyl sulfide, dimethyltrisulfide	Sulfide/putrid
	Acinetobacter	Aliphatic alcohols and aldehydes, aromatic hydrocarbons	Sour
	Serratia liquefaciens	2-Methylbutanol, 3-methylbutanol, hydrogen sulfide, methanethiol, dimethyl disulfide, dimethyl trisulfide, methyl thioacetate	Putrid

	Organism	Compounds	Descriptor
Chicken	Clostridium ethertheticum	Hydrogen sulfide, methanethiol, butanol, butyl butanoate, butyl acetate, butanoic acid, dimethyl sulfate	Sulfurous/fruity
	Pseudomonas fluorescens	Dimethyl sulfide, dimethyl disulfide, methyl thioacetate	Faecal
	P. putida	Methanethiol, carbon disulfide, methyl thioacetate, dimethyl disulfide, ethyl methyl disulfide, dimethyl trisulfide	Putrid
	S. putrefaciens	Hydrogen sulfide, dimethyl disulfide, dimethyl trisulfide	Sulfide/putrid
	B. thermosphacta	3-Methylbutanol, 2-methylpropanol, 3-methylbutanol, 2-hexanone, 3-methylbutanoic acid, methyl 3-hydroxy-3-pentenoate	Buttery/cheesy
	Serratia liquefaciens	2-Methylbutanol, 3-methylbutanol, hydrogen sulfide, methanethiol, dimethyl disulfide, dimethyl trisulfide, methyl thioacetate	Putrid
Eggs	Pseudomas fluorescens	Dimethyl sulfide, dimethyl disulfide, methyl thioacetate	Faecal
	P. perolens P. taetrolens	2-Methoxy-3-isopropyl pyrazine	Musty/potato

Table 5.2 Continued

Food	Microorganisms	Compounds produced	Off-flavour
Fish	*S. putrefaciens, Photobacterium phosphoreum*, Vibrionaceae	Trimethylamine	Ammoniacal/fishy
	S. putrefaciens, P. fragi, P. fluorescens, Alcaligenes/Achromobacter spp. *Photobacterium phosphoreum*, Vibrionaceae	Methanethiol, hydrogen sulfide, dimethyl disulfide, dimethyl trisulfide	Rotten/hydrogen sulfide
	P. fragi	Ethyl acetate, ethyl butanoate, ethyl hexanoate	Fruity
	P. perolens	2-Methoxy-3-isopropylpyrazine	Musty/potato
	Oscillatoria spp.	Geosmin	Earthy
	Lyngbya spp.		
	Anabaena spp. *Microcystis* spp. *Oscillatoria* spp.	2-Methylisoborneol	Muddy
Fruit, vegetables and cereals	*Lactobacillus* spp., *Leuconostoc* spp., *Lb. plantarum*	Diacetyl, acetoin, 2,3-dihydroxybutane	Buttermilk
	Penicillium italicum, P. digitatum	4-Vinylguaiacol	Rotten fruit
	Gram-positive diplococci	Butanoic acid, C_5 to C_9 alkanoic acids, methyl C_4, C_6, C_8 alkanoates	Butyric/acrid
	Penicillium expansum	Geosmin	Earthy
	Alicyclobacillus acidoterrestris	Guaiacol	Smoky
	Penicillium spp., *Aspergillus* spp., *Eurotium* spp., *Paecilomyces variotii*, *Fusarium oxysporum*	2,4,6-Trichloroanisole	Musty/mouldy
	Eurotium spp., *Penicillium citrinum, P. solitum*	Odd-numbered methyl ketones (C_5 to C_{11})	Ketonic/perfume-like
	Actinomycetes, *Streptomyces griseus*	Geosmin, 2-methylisoborneol	Earthy
	Clostridium spp., *C. scatologenes*	Skatole, indole, *p*-cresol	Faecal

of gaseous compositions. Aerobic bacteria must have oxygen for growth whereas anaerobic bacteria will only grow in oxygen-depleted atmospheres. However, facultatively anaerobic bacteria will grow both in the presence and absence of oxygen. Consequently, foods such as meat, poultry, fish and dairy products are particularly susceptible to bacterial spoilage. Such spoilage is usually characterised by growth to high numbers from an initially low number of bacteria. Under aerobic conditions the most common spoilage bacteria are Gram negative aerobic rods such as *Pseudomonas* spp. and *Shewanella* and facultatively anaerobic Gram negative rods such as *Aeromonas* and *Photobacterium*. At chill temperatures *Pseudomonas* spp. usually predominate whereas at higher temperatures vibrionaceae and other enterobacteriaceae are the dominant bacteria. However, under anaerobic conditions such as are present in vacuum-packaged products, non-spore-forming lactic acid bacteria including *Lactobacillus* and *Brochothrix* can predominate. Spore-forming bacteria usually grow more slowly than Gram negative bacteria but can become important contaminants in foods that have received heat processing. Many of these bacteria have the ability to be thermotolerant. Not all of these bacteria will produce off-flavours, their presence will be indicated by other forms of food spoilage such as discoloration and liquefaction. However, off-flavours can be produced in foods by many species of aerobic, anaerobic and spore-forming bacteria provided the conditions of processing and storage are favourable to their growth. The following section will identify some of these bacteria, together with the off-flavour compounds they can produce and the possible precursors of such compounds.

5.3 Aerobic bacteria

5.3.1 *Acinetobacter*

All strains of *Acinetobacter* grow between 20 and 30°C with most strains having temperature optima between 33 and 35°C. These bacteria have a slightly acid pH optimum for growth between 5.5 and 6.0. *Acinetobacter* occur naturally in soil and water and are also present in sewage.[3] At present all strains of *Acinetobacter* are considered to be members of a single species, *Acinetobacter calcoaceticus*.[3]

Strains of *Aci. calcoaceticus* have been identified as a cause of a sour off-odour in spoiled ground beef.[4] The isolated *Acinetobacter* were divided into two groups, FI (non-fermatative) and FII (fermatative). The FI isolate when grown on ground beef produced 22 volatile compounds, principally aliphatic alcohols, esters and hydrocarbons and aromatic hydrocarbons, whereas the FII isolate produced only 12 compounds, mainly aliphatic aldehydes and aromatic hydrocarbons. Low molecular weight fatty acids, normally associated with sour odours, were not identified. The compounds possibly associated with the sour odour observed in the beef inoculated with the FI isolate were 2,3-pentadiene, acetaldehyde, hexanol, 2-methylbutan-2-one, dimethyl sulfide, ethyl propanoate, dimethyl benzene and diethyl benzene. For the FII isolate the possible odorous compounds were 4-methylpentanal, pentanal, 2-pentanone, ethyl acetate, toluene and dimethyl benzene. Accordingly, the majority of compounds formed in beef inoculated with these strains of *Aci. calcoaceticus* were derived from the decomposition of lipids and to a less extent the sulfur amino acids. However, the isolate FII appears to be less active with fewer compounds produced, and only trace quantities of dimethyl sulfide formed.

5.3.2 *Alcaligenes* (*Achromobacter*)

The genera *Alcaligenes* and *Achromobacter* were long considered the repositories for different types of bacteria, most of which were poorly characterised. Consequently, both names are used randomly throughout the literature. *Alcaligenes* are obligately aerobic but some strains are capable of anaerobic respiration in the presence of nitrate or nitrite. The optimum temperature for growth of this genera is 20 to 37°C. The natural source of these bacteria are water, soil and faeces.[3]

Species of *Alcaligenes* (*Achromobacter*) isolated from Dungeness crab were found to produce characteristic spoilage odours and flavours when inoculated into sterile fish muscle.[5] The compounds produced were methanethiol, dimethyl disulfide, 1-penten-3-ol, 3-methylbutanal and trimethylamine. These were formed in fish muscle at pH 6.4 to 6.7 when incubated at 1 to 2°C for 12 and 27 days. The principal source of these compounds would appear to be amino acids and trimethylamine oxide. Some lipid degradation also appears to occur in the presence of these bacteria.

5.3.3 *Alteromonas*

Species of the *Alteromonas* genera all require a sea water base for growth and are common inhabitants of coastal waters and open oceans.[3] All species will grow at 20°C but the range of growth extends from 4 to 40°C. Of these bacteria, one species diversely named, *Alteromonas putrefaciens*, *Achromobacter putrefaciens* and *Pseudomonas putrefaciens*, contains strains of major importance in seafood spoilage. This species was recently transferred to its own genus, *Shewanella*, and it is under this name that its role in food spoilage will be discussed.

5.3.4 *Alicyclobacillus*

The genus *Alicyclobacillus* was only described in 1992.[6] All *Alicyclobacillus* are acid dependent, spore-forming, thermotolerant, aerobic bacteria that naturally occur in soils and hot springs. Such bacteria can grow between 20 and 58°C and at pHs between 2.5 and 5.5. As yet only one species of this genus, *Alicyclobacillus acidoterrestris* is known to produce off-flavours in foods.

Alicyclobacillus acidoterrestris

Alicyclobacillus acidoterrestris was first associated with the presence of off-flavours in fruit juice in 1984.[6] Since then this bacterium has been identified in spoiled fruit juices in many parts of the world. Significantly, only small numbers of bacteria (1×10^5) are necessary to produce spoilage. The off-flavour is variously described as disinfectant, smoky and phenolic and the compounds identified in the off-flavoured juices are guaiacol,[7] 2,6-dibromophenol[8,9] and 2,6-dichlorophenol.[9] It is accepted that guaiacol is the metabolite responsible for the smoky off-flavour in the spoiled juice, but evidence supporting halophenols as the cause of the disinfectant off-flavour has been disputed. However, in studies it has been shown that strains of *Ali. acidoterrestris* isolated from Australian fruit juice produced 2,6-dibromophenol and 2,6-dichlorophenol as major halogenation products.[9] It has been suggested that these strains contain a haloperoxidase that can oxidise halogen ions to active halogens that react with suitable phenolic precursors in the fruit juice.[9] The likely precursor for guaiacol is ferulic acid, a common component of some fruit juices. Several microorganisms, including the yeast *Rhodotorula ruba* are known to convert this acid by β-oxidation and subsequent decarboxylation of the resultant vanillic acid to guaiacol.[10] *Ali. acidoterrestris* possibly contain similar enzyme systems.

5.3.5 *Moraxella*

All strains of *Moraxella* grow between 33 and 35°C, however, one species can grow at 5°C.[3] These bacteria are parasitic on the mucous membranes of man and other warm-blooded animals including cattle, sheep, goats and pigs. Species of *Moraxella* have been shown to cause a pronounced estery and decayed-vegetable-like odour in spoiled ground beef.[4] Some 15 compounds were identified in the headspace above beef inoculated with these species and stored at 10°C for 5 days. The main compounds were a dimethyl benzene, ethyl acetate, dimethyl sulfide and pentane. Other compounds identified included dimethyl disulfide, methyl acetate, ethyl propanoate and propyl acetate. The four esters identified probably account for the estery odour of the spoiled beef, whereas the sulfides possibly account for the decayed-vegetable odour. The presence of significant quantities of dimethyl sulfide indicates that these bacteria can metabolise methionine and possibly also cysteine.

5.3.6 *Pseudomonas*

Pseudomonas are one of the most complex groups of Gram negative bacteria with phenotype similarities to many other genera.[3] At one point this single genus contained over 100 species, which have been assigned to nine different genera including *Pseudomonas*. Even so, *Pseudomonas* probably account for more off-flavour problems in foods than any other bacteria. The off-flavours caused by the presence of five species of *Pseudomonas* will be discussed in the following section.

Pseudomonas fluorescens
Pseudomonas fluorescens is a very large species with at least five biovariants of which biovar III has 68 strains and biovar V 89 strains. The optimum temperature for growth of these bacteria is 25 to 30°C, although many strains will grow at 4°C. The natural habitat for these bacteria is soil, water, sewage and faeces.[3] They are commonly associated with the spoilage of eggs, cured meats, fish and milk. The presence of *Pse. fluorescens* on certain foods is indicated by a blue green fluorescence and putrid odours.[3]

Unclean, bitter and putrid flavours are known to occur in pasteurized milk as a result of post-pasteurization contamination with psychrotrophic bacteria including *Pse. fluorescens*. Proteolytic strains of this bacterium degrade caseins and whey products to produce bitter flavours. Peptides are suspected to be the cause of the bitterness but attempts to identify these compounds have been unsuccessful.[11]

Pse. fluorescens also occurs on raw fish where it has been associated with low temperature spoilage of such products. Inoculation of fish muscle with this bacterium followed by incubation at 0°C for 32 days resulted in putrid odours. Compounds identified were methanethiol and dimethyl disulfide. It is likely that these compounds were the same as those responsible for putrid off-flavours in pasteurized milk contaminated with *Pseudomonas* species.[12]

Pse. fluorescens is a frequent microbial contaminant of chicken meat. Strains of this bacterium grown on sterilized chicken breasts for 5 days at 10°C led to the production of eight compounds including dimethyl sulfide, dimethyl disulfide and methyl thioacetate.[13] In further studies some 19 compounds were identified when chicken breasts were inoculated with *Pse. fluorescens* and incubated for 14 days at 2 and 6°C.[14] New compounds identified included methanethiol, ethyl methyl disulfide, 2- and 3-pentanone, 2-heptanone, 3- and 4-octanone, 2-nonanone and nonanol. All of the alkanones and the alkanol would appear to be derived from the decomposition of the lipid components, and the sulfur compounds from the metabolism of methionine, cysteine and cystine. A further 14 compounds were identified when carcasses inoculated with this bacterium were incubated for 4 to 7 days at 4°C. Included among the new compounds were hydrogen sulfide, dimethyl tetrasulfide, indole and 1-octen-3-ol.[15] Indole with its faecal-like odour could be expected to contribute to the offensive odour of the spoiled carcasses. This compound would be derived from the decomposition of the amino acid tryptophan.

The role of *Pse. fluorescens* in the spoilage of ground beef has also been studied.[4] In beef this bacterium can produce a decayed straw-like odour after 5 days storage at 10°C. Some 12 compounds were identified of which the major compounds were dimethyl disulfide and 2-butanone. Ammonia and trimethylamine were also detected in the spoiled samples. This result is interesting as neither of these nitrogen compounds was found when raw fish was inoculated with strains of this bacterium.[5]

Heat-resistant lipases from psychrotrophic strains of *Pseudomonas* spp. contribute to the off-flavour of rancid milk by the release of short- and medium-chain length fatty acids from the milk triglycerides.[16] In particular, the lipase isolated from *Pse. fluorescens* yields butanoic, hexanoic, octanoic and decanoic acids when added to milk-fat slurries.[17] Such compounds are associated with the bitter or soapy-like off-flavour associated with rancid milk.

Pseudomonas fragi

Pseudomonas fragi is a major cause of microbiologically derived off-flavours in meat and dairy products. The natural habitat of this bacterium is soil and water but it is frequently found in food processing areas.[3] Its optimum temperature of growth is 10 to 30°C. *Pse. fragi* is a non-fluorescent species of *Pseudomonas*.[3]

Pse. fragi has been associated with the spoilage of dairy products as a result of post-pasteurization contamination. Such spoilage is characterised by fruity odours and a strawberry-like off-flavour.[18] The compounds responsible for the fruity odour are ethyl acetate, ethyl butanoate and ethyl hexanoate.[19] *Pse. fragi* is a strongly lipolytic bacterium that removes fatty acids preferentially from the 1- and 3-positions of triglycerides.[20] Butanoic and hexanoic acids are esterified in the 3-position of milk triglycerides and are the major products of lipolysis by this bacterium.[21] *Pse. fragi* also produces ethanol, presumably from sugar fermentation,[19] and excessive production of this compound in stored milk can be responsible for enhanced esterification of some free fatty acids.[16] The esterification of butanoic and hexanoic acids by ethanol is facilitated by the presence of an esterase in the cell extracts of some *Pseudomonas* spp.[22,23]

Strains of *Pse. fragi* have been identified as frequent microbial contaminants of commercial fish fillets, where under chilled storage they can produce fruity and onion-like odours.[24] Fruity odours were observed during the early stages of spoilage of inoculated fillets and corresponded to the formation of ethyl acetate, ethyl butanoate, ethyl hexanoate and ethanol.[25] An intense sulfide or onion-like odour developed with continued incubation and corresponded to the formation of methanethiol, dimethyl sulfide and dimethyl disulfide. The formation of the sulfur compounds shows that this bacterium, like *Pse. fluorescens,* can readily metabolise methionine and possibly also cysteine and cystine.

Pse. fragi is a frequent contaminant of chicken breasts[13] and poultry carcasses.[15] In studies where whole poultry carcasses were inoculated with three different strains of this bacterium and incubated for 4 and 7 days at

4°C, a total of 62 compounds were identified.[15] These metabolites included 27 aliphatic esters and 17 sulfur compounds. Of the esters, those produced in greatest quantity were ethyl acetate, ethyl butanoate, ethyl hexanoate, ethyl octanoate, ethyl 2-methylpropanoate, ethyl 2-methylbutanoate, ethyl 3-methylbutanoate, methyl 2-methylenebutanoate and ethyl 2-methyl-3-butanoate. The acid component of these esters was mainly derived from two sources, from the lipolysis of triglycerides and from the decomposition of the amino acids, valine, leucine and isoleucine. The dominant alkyl component of these esters were ethyl, methyl and propyl. The major sulfur compounds produced by these strains were methanethiol, dimethyl sulfide, dimethyl disulfide, dimethyl trisulfide and dimethyl tetrasulfide.

Pse. fragi has been associated with the development of sweet, putrid and fruity odours in refrigerated normal and high pH beef.[26] Some 21 compounds including 11 esters and six sulfur compounds were formed when beef inoculated with this bacterium was stored for 6 to 8 days at 5°C. Ethyl esters of short chain fatty acids and deaminated amino acids dominated the ester fraction and all the sulfur compounds were derivatives of methanethiol. The role of *Pse. fragi* in the development of sulfury/putrid and ammoniacal odours in ground beef has also been studied.[4] Some 31 compounds were identified from inoculated samples of which aliphatic and aromatic hydrocarbons were qualitatively the major metabolites. Quantitatively, dimethyl disulfide, methanethiol and dimethyl sulfide were the dominant metabolites. Aliphatic esters were only minor components. Accordingly, the volatile compounds identified in this study differed appreciably from those found by other workers studying the metabolites of this bacterium.[15,26]

Pseudomonas perolens (Achromobacter perolens)
Achromobacter perolens was first described as the bacterium responsible for a penetrating musty odour in eggs in 1927.[27] The species was renamed *Pseudomonas perolens* in 1950[28] and this name was retained until 1974 when it disappeared from the Register. However, the name *Pseudomonas perolens* has continued to be used in some literature.

Pseudomonas perolens has been associated with the development of musty and potato-like odours in chilled fillets of cod, haddock and flounder, and steaks of halibut that had been stored under commercial conditions.[29] Compounds formed when this bacterium was grown on fish muscle incubated at 5, 15 and 25°C, included methanethiol, dimethyl disulfide, dimethyl trisulfide, 3-methylbutanol, 2-butanone and 2-methoxy-3-isopropylpyrazine.[30] 2-Methoxy-3-*sec*-butylpyrazine was tentatively identified. The musty and potato-like odour of the spoiled fish was attributed to the presence of 2-methoxy-3-isopropylpyrazine and possibly also 2-methoxy-3-*sec*-butylpyrazine. Metabolism of the amino acids methionine, cysteine and cystine was considered to be the likely source of the sulfur compounds.[30]

However, it was not until 1991 that a satisfactory biosynthetic pathway for the formation of 2-methoxy-3-isopropylpyrazine was achieved.[31] Based on labelling experiments it was shown that this compound could be produced when *Pse. perolens* was grown in a medium containing pyruvate as the sole source of carbon. Consequently, it was suggested that endogenous valine, glycine and methionine were the likely precursors of 2-methoxy-3-isopropylpyrazine produced by this bacterium.

Pseudomonas putida
Pseudomonas putida exists as two biovars, A (103 strains) and B (9 strains). It is a fluorescent species and is very closely related to *Pse. fluorescens*. This bacterium has an optimum temperature for growth of 25 to 30°C, although some strains will grow at 4°C. The natural habitat for this bacterium is soil, water and putrefying materials.[3]

Pse. putida has long been associated with the spoilage of chicken meat.[13,14] Some 19 compounds were isolated from chicken breasts inoculated with this bacterium and incubated for 14 days at 2 to 6°C. The majority of compounds were either alkanones (eight) or sulfur compounds (six). The compounds believed to be responsible for the putrid odour were methanethiol, carbon disulfide, methyl thioacetate, dimethyl disulfide, ethyl methyl disulfide and dimethyl trisulfide. The number of compounds formed increased to 37 when chicken carcasses inoculated with *Pse. putida* were incubated for 7 days at 4°C.[15] The major compounds formed were 2-hexanone, 2-heptanone, 2-nonanone, 5-nonen-2-one, 2-pentanone, 2-octanone, 2-pentanol, 2-heptanol, 2-nonanol, hydrogen sulfide and methanethiol. The large number of both even and odd carbon number aliphatic ketones produced would suggest that this bacterium is capable of conventional as well as abortive β-oxidation of fatty acids. Furthermore, the presence of the corresponding 2-alkanols indicates that the bacterium is also capable of ketone reduction, presumably with the assistance of an alcohol dehydrogenase.

Pse. putida has been associated with the spoilage of ground beef where it produces a sour odour.[4] Some 15 compounds were identified, principally aliphatic and aromatic hydrocarbons. Dimethyl disulfide was the only sulfur compound identified.

Pse. putida has been identified as one of the causes of a fruity odour and sour rancid off-flavour in pasteurized milk.[32] Compounds responsible for the odour were ethyl butanoate, ethyl hexanoate, ethyl octanoate and ethyl decanoate and those responsible for the rancid flavour were octanoic acid, decanoic acid and dodecanoic acid. Inoculation of milk with an isolate of *Pse. putida* gave five even carbon number fatty acids (C_4 to C_{12}) together with the ethyl esters of the C_4, C_6 and C_8 acids. Accordingly, *Pse. putida* contain enzyme systems that are capable of lipolysis and esterification similar to those systems associated with *Pse. fluorescens* and *Pse. fragi*.

Pseudomonas taetrolens (Pseudomonas graveolens)
Pseudomonas graveolens was first isolated from eggs with a musty and potato-like odour in 1932.[33] This species was renamed *Pseudomonas taetrolens* and as such was identified as the cause of potato or cellar-like odours in lamb carcasses[34] and in milk.[35] Compounds formed when this bacterium was grown in skim milk cultures included 2,5-dimethylpyrazine and 2-methoxy-3-isopropylpyrazine.[35] A possible biosynthetic pathway for the formation of 2,5-dimethylpyrazine was proposed.[35] Growth of an isolate of *Pse. taetrolens* in a medium containing yeast, glucose and Bacto-tryptone™ led to the biosynthesis of five volatile compounds, 2-methoxy-3-isopropylpyrazine, methylpyrazine, 2,5-dimethylpyrazine, 2,3,5-trimethylpyrazine and 2,5-dimethyl-3-ethylpyrazine.[36] However, when grown in a minimal basal salts medium supplemented with either leucine or valine, the yield of the 2-methoxy-3-isopropylpyrazine was greatly increased.[36] The biosynthesis of this compound by *Pse. taetrolens* would appear to be the same or similar to that used by *Pse. perolens*.

5.3.7 *Shewanella putrefaciens* (*Pseudomonas putrefaciens, Altermonas putrefaciens, Achromobacter putrefaciens*)

Since the early 1940s the species *Shewanella putrefaciens* has had four names and articles describing the behaviour of this bacterium appear under all four classifications. This bacterium is an important spoilage organism of food principally because of its ability to spoil low temperature, high pH, high protein foods. All strains of *She. putrefaciens* are psychrotrophic, growing at 4°C and in many cases at 0°C, but rarely grow at 37°C. The natural habitat of this bacterium is water, soil and marine sediment. However, although associated with marine environments it does not tolerate high concentrations (10%) of salt.[3]

She. putrefaciens has long been associated with the spoilage of raw fish and its presence is characterised by ammoniacal, rotten and hydrogen sulfide-type odours.[37,38] Inoculation of fish muscle with *She. putrefaciens* followed by incubation for 12 to 27 days at 1 to 2°C, led to the formation of six metabolites, methanethiol, dimethyl disulfide, dimethyl trisulfide, 3-methylbutanol, 1-penten-3-ol and trimethylamine.[5] Reduction of trimethylamine oxide to trimethylamine was most pronounced between 4 and 8 days incubation. Accordingly, this bacterium, like several others, utilises trimethylamine oxide as part of its anaerobic respiration. *She. putrefaciens* is also a major producer of hydrogen sulfide.[39] Methanethiol, hydrogen sulfide and the other sulfur compounds are all metabolites of methionine, cysteine, cystine and possibly glutathione.[5]

She. putrefaciens has also been associated with the spoilage of normal and high pH beef where it can produce an egg-like or putrid odour.[26] The four sulfur compounds previously associated with the spoilage of fish were identified in inoculated high pH beef together with methyl thioacetate and bis-(methylthio)methane. Bis-(methylthio)methane has an intense garlic-like odour.

This bacterium has also been associated with the spoilage of chicken meat.[13,15] In the more recent studies[15] poultry carcasses inoculated with three strains of *She. putrefaciens* were incubated for 7 days at 4 to 7°C. Some 27 compounds were identified, of which the major compounds were hydrogen sulfide, dimethyl disulfide and dimethyl trisulfide. Another 13 sulfur compounds were identified, principally sulfides and thioesters. Interestingly, bis-(methylthio)methane was not identified among these metabolites.

5.4 Facultative anaerobic bacteria

5.4.1 *Bacillus*

Species of the genus *Bacillus* are endospores that are very resistant to adverse conditions. They are aerobic or facultatively anaerobic and will grow at pHs between 5.5 and 8.5.[40] Volatile metabolites produced by these bacteria include 2,3-butanedione and butanoic acid and these compounds have been used to identify the presence of such spore formers in canned products.[41]

Bacillus subtilis
Bacillus subtilis is the causative agent of ropy (slimy) bread and has been associated with the development of a pungent odour in heat-treated and gamma-irradiated coconut. Compounds present in the off-flavoured coconut were hexanoic acid, octanoic acid, decanoic acid, dodecanoic acid, acetoin, 2,3-butanedione, 2,3-butanediol, 2,3,5-trimethylpyrazine and 2,3,5,6-tetramethylpyrazine.[42] The fatty acids are well known metabolites of this bacterium, which possesses strong lypolytic activity.[43] The compound principally responsible for the off-odour, 2,3,5,6-tetramethylpyrazine had been previously isolated from cultures of *Bac. subtilis* grown on media containing glucose and asparagine.[44]

5.4.2 *Brochothrix*

The genus *Brochothrix* contains only two type species, *Brochothrix thermosphacta* and *Brochothrix campestris*. *Bro. thermosphacta* is commonly found on processed meat and fresh and processed meat that have been stored in low oxygen permeable films at refrigeration temperatures. *Bro. campestris* is indigenous to the farm environment.[40]

Brochothrix thermosphacta
Growth of *Bro. thermosphacta* occurs between 0 and 30°C and at pHs from 5 to 9. All strains of this bacterium will grow in the presence of 6.5% salt and the minimum a_w for growth is 0.94.[40]

Under anaerobic conditions *Bro. thermosphacta* can produce lactic acid as the major end product of glucose fermentation. Some short chain fatty acids are also formed under these conditions. Accordingly, beef contaminated with this bacterium can develop sickly sweet or sickly sour odours at low temperatures (1°C). Compounds associated with spoiled normal pH beef include acetoin and acetic acid together with smaller quantities of propanoic acid, butanoic acid, 2-methylpropanoic acid and 3-methylbutanoic acid.[45] In spoiled high pH beef the amount of acetoin found was greatly reduced but the levels of the fatty acids were about the same. In subsequent studies it was shown that acetoin and acetic acid were derived from glucose, whereas the branched chain acids, 2-methylpropanoic acid, 3-methylbutanoic acid and 2-methylbutanoic acid were metabolites of valine, leucine and isoleucine, respectively.[46] However, in the absence of amino acids, all of the end products including the branched chain acids could be derived from glucose. Further studies led to the identification of 2-methylpropanol, 2-methylbutanol, 3-methylbutanol and 3-methylbutanal. These branched chain alcohols and aldehyde were all secondary metabolites of the corresponding acids.[26]

Bro. thermosphacta has also been associated with the spoilage of chicken meat.[13,15] Forty four compounds were identified when poultry carcasses were inoculated with three strains of this bacterium and the carcasses incubated for 4 and 7 days at 4°C.[15] Major variations in which compounds were produced occurred across the three strains. The culture that produced the greatest number of compounds (39) was subsequently shown to be contaminated with native pseudomonad flora. Major compounds produced by the other two strains were 3-methylbutanol, 2-methylpropanol, 3-methylbutanal, 2-hexanone, 3-methylbutanoic acid and methyl 3-hydroxy-3-pentenoate. These results confirm those found previously with beef that *Bro. thermosphacta* does not normally metabolise the sulfur amino acids.[26,45]

5.4.3 *Serratia*
The genus *Serratia* belongs to the family Enterobacteriaceae. Most species grow at temperatures between 10 and 36°C (some at 4 or 5°C), at pH 5 to 9, and in the presence of 0 to 4% sodium chloride.[3] Cultures of *Serratia* species can produce two kinds of odours, a fishy-urinary and a musty potato-like odour. The musty odour is produced by *Serratia odorifera*, *Serratia ficaria* and *Serratia rubidaea*. All other species produce the fishy-urinary odour.[3]

Serratia liquefaciens
Serratia liquefaciens is the most prevalent *Serratia* species in the natural environment (plants, digestive tract of rodents) and is occasionally

encountered as an opportunistic pathogen. It will grow over the temperature range 4 to 37°C and at pH 5.5 to 6.8. It is a frequent contaminant of meat and poultry where it is associated with putrid and rotten egg-like odours.[3]

Inoculation of normal pH (5.5 to 5.8) and high pH (6.6 to 6.8) beef with *Ser. liquefaciens* followed by incubation for 4 to 7 days at 5°C led to the production of 10 compounds at normal pH and eight at high pH.[26] Compounds common at both pHs were 2-methylbutanol, 3-methylbutanol, hydrogen sulfide, methanethiol, dimethyl disulfide, dimethyl trisulfide and methyl thioacetate. Additional compounds found in the normal pH beef were 2,3-butanediol, 2,3-butanedione, and acetoin, whereas at the high pH the only additional compound was bis-(methylthio)methane. Both the normal and high pH beef cultures were described as possessing a rotten egg-like odour.[26]

Two strains of *Ser. liquefaciens* were used to study the role of this bacterium in the spoilage of refrigerated poultry carcasses.[15] Twenty compounds were identified, of which hydrogen sulfide, 3-methylbutanol and 2-hexanone were the major compounds. Of the other compounds identified, eight were aliphatic alcohols and five were 2-alkanones. Accordingly, only five of the compounds found in this study had been identified previously in beef.[26]

Serratia ficaria, Serratia marcescens, Serratia odorifera
and Serratia rubidaea
Broth and agar cultures of four species of *Serratia, Serratia ficaria, Serratia marcescens, Serratia odorifera* and *Serratia rubidaea* are all known to possess intense potato or fish-like odours.[47] Cultures of *Ser. ficaria* have a potato-like odour and the major compound produced by this bacterium is 2-methoxy-2-isopropyl-5-methylpyrazine together with a small quantity of 2-methoxy-3-*sec*-butyl-5(6)-methylpyrazine. Cultures of *Ser. marcescens* produces 2,3,5-trimethylpyrazine, those of *Ser. odorifera* 2-methoxy-3-isopropylpyrazine and a small quantity of 2-methoxy-3-isobutyl-6-methylpyrazine, and those of *Ser. rubidaea* 2-methoxy-3-isopropyl-5-methylpyrazine, 2-methoxy-3-*sec*-butylpyrazine, 2-ethyl-6-methylpyrazine and 2,3,5-trimethylpyrazine. Metabolic pathways to the different 2-methoxyalkylpyrazines produced by these bacteria are probably similar to those proposed for the biosynthesis of 2-methoxy-3-isopropylpyrazine by *Pse. perolens* and *Pse. taetrolens*.

5.4.4 Rahnella

The genus *Rahnella* is a member of the family Enterobacteraceae and at present there is only one species, *Rahnella aquatilis*.[3] Its natural habitat is freshwater but it is also found in wort and a range of dairy products. This bacterium will grow at temperatures between 4 and 37°C and at a pH as low as 5.[3]

The psychrotrophic bacterium *Rah. aquatilis* has been associated with the development of a smoky/phenolic odour and flavour in chocolate milk.[48] The compound responsible was a guaiacol, and growth of *Rah. aquatilis* in UHT chocolate milk and UHT white milk led to the formation of this compound but only in the flavoured milk. Known precursors of guaiacol are vanillin, vanillic acid and ferulic acid.[48] Studies showed that vanillin could not be metabolised to guaiacol by this bacterium. It was suggested that the likely precursor of guaiacol in chocolate milk was vanillic acid, an oxidation product of vanillin.[48]

5.4.5 *Yersinia*
The genus *Yersinia* is a member of the family Enterobacteriaceae.[3]

Yersinia intermedia
The major source of *Yersinia intermedia* is fresh water but it also occurs as a contaminant in fish and other foods.[3] It is a psychrotrophic bacterium that has an optimum temperature for growth at about 23°C. This bacterium has been associated with the development of a fruity and soapy flavour in pasteurized milk.[32] Compounds isolated from the spoiled milk included ethyl butanoate, ethyl hexanoate, ethyl octanoate, ethyl decanoate, octanoic acid, decanoic acid and dodecanoic acid. Growth of an isolate of *Yer. intermedia* in UHT milk resulted in the development of the same fruity odour. Compounds found in the inoculated culture included all of the compounds found in the spoiled milk together with butanoic acid and hexanoic acid.[32] In this study *Yer. intermedia* displayed strong lipolytic activity and readily liberated short chain (C_4 to C_{12}) fatty acids from milk triglycerides. This bacterium also contains a dehydrogenase to reduce acetaldehyde to ethanol and an esterase that is capable of esterifying the C_4 to C_8 fatty acids with ethanol.[32]

5.5 Anaerobic bacteria

5.5.1 *Clostridium*
Most *Clostridium* species are obligately anaerobic although tolerance to oxygen varies widely within the species. Growth is most rapid at pH 6.5 to 7 and at temperatures between 30 and 37°C. The range of temperature for optimum growth is from 15 to 69°C. Clostridia are ubiquitous, commonly found in soil, sewage, marine sediments, decaying vegetation, animal and plant products.[40]

Clostridium estertheticum

Clostridium estertheticum was first identified as a cause of spoilage of vacuum-packed beef in 1989, but was not fully described until 3 years later. It was isolated from spoiled chill-stored normal pH beef that had an odour described as sulfurous, fruity, solvent-like and strong cheese.[49] Hydrogen and carbon dioxide were the major headspace gases. Some 20 compounds were identified, of which the major compounds were hydrogen sulfide, methanethiol, butanol, butyl butanoate, butyl acetate, butanoic acid and dimethyl sulfide. In some samples ethyl butanoate and butyl formate were also produced in significant quantities. Butanol and butanoic acid are characteristic end products of the metabolism of glucose by saccharolytic clostridia. In addition, this bacterium contains an esterase capable of esterifying some fatty acids with butanol. It can also metabolise the sulfur amino acids.

Clostridium scatologenes

Clostridium scatologenes has an optimum temperature for growth of between 30 and 37°C and will grow at 25°C but not at 45°C. Its natural habitat is soil.[40]

Clo. scatologenes has been associated with a faecal or pigsty-like off-flavour and odour in frozen potato chips (French fries).[50] The compounds identified in the spoiled chips and potato tubers stored in the processing plant were skatole, indole and *p*-cresol. The tubers were found to be contaminated with the facultatively anaerobic bacteria *Erwinia carotovora*, *Erwinia chrysanthemi* and the anaerobic bacterium *Clo. scatologenes*. Of these bacteria only *Clo. scatologenes* was known to produce skatole, indole[51] and *p*-cresol.[52]

So-called 'boar taint' is attributed by some to the presence of skatole in the back fat of susceptible animals. Skatole is produced in the hindgut of pigs by microbial degradation of tryptophan originating from dietary and endogenous protein. The bacteria found in the gut of pigs that can produce skatole include a strain of *Lactobacillus helveticus* and *Clo. scatologenes*.[53] Studies showed that whereas *Clo. scatologenes* degraded tryptophan directly to skatole, the strain of *Lac. helveticus* first required the tryptophan to be degraded to indole-3-acetic acid before it could produce skatole.[53]

5.6 Actinomycetes

Actinomycetes are widely distributed in nature and account for a large part of the normal microbiological population in soils, and lake and river muds.[54] They are also well known for their production of intense penetrating odours, in particular the earthy odour of freshly tilled soil and the muddy odour of some muds. Some ten species of actinomycetes are known to produce geosmin, the earthy odour of soil, seven species also produce 2-methylisoborneol, the muddy odour of mud, and at least one species produces 2-methoxy-3-isopropylpyrazine the musty potato-like odour of cellars.[54] Actinomycetes have been associated with earthy and muddy off-flavours in fish, wine, mushrooms and cereal grains; however, there are only a few incidences where the actinomycetes involved in the spoilage of food have been identified. One such incidence is discussed below.

5.6.1 *Streptomyces griseus*
Earthy-musty off-flavours have been observed in some cereal products, and studies have shown that actinomycetes capable of producing geosmin and 2-methylisoborneol can grow in grain bread containing 45 to 50% moisture[55] and on wheat grain with 16 to 18% moisture.[56] In 1991 a bulk shipment of wheat flour with an earthy odour was found to be contaminated with geosmin and to a lesser extent 2-methylisoborneol.[57] Microbiological examination of the flour resulted in the identification of *Streptomyces griseus*. Growth of this organism on wheat grains at 50% moisture gave geosmin after only 2 days incubation. Interestingly where bread was inoculated with a different strain of *Str. griseus*, the ratio of 2-methylisoborneol to geosmin was 10:1.[55] In this study 3-methylbutanol and dimethyl trisulfide were also identified as metabolites.[55]

5.7 Fungi

Fungi responsible for food spoilage can grow at a_w values from 0.62 to 0.99, pH values from 2 to 10, temperatures from −3 to 50°C and, like some bacteria, certain fungi can grow in the absence of oxygen. However, most food spoilage fungi have an absolute requirement for oxygen, which can be as low as 1%. Most foods have the nutrient status to support the growth of fungi; but in general fungi are best suited to substrates high in carbohydrates, whereas bacteria are more likely to spoil proteinaceous foods.[58] Xerophilic fungi are capable of growth at reduced a_w values (below 0.85) and as a consequence can cause spoilage problems in a wide range of foods and other commodities including fibreboard packaging materials. Non-discriminating fungi and those that are moderate xerophiles such as species of *Eurotium*, *Aspergillus* and *Penicillium* are wide ranging in their habitat and are responsible for most fungal spoilage in stored grains, spices, dried fruits, nuts and oil seeds.

In foods of somewhat higher a_w such as processed dairy products including cheddar cheese and cream cheese, spoilage can be caused by species of *Cladosporium*, *Penicillium* and *Phoma*. Most fungi responsible for food spoilage prefer to grow at temperatures between 5 and 37°C. However, there are some species such as *Aspergillus flavus* and *Aspergillus niger*, which grow at both moderate and high temperatures (between 8 and 45°C), that are amongst the most destructive fungi known.[58]

Off-flavours of fungal origin can be produced in foods by the metabolism of suitable natural precursors and additives. They can also be absorbed from packaging materials and litter contaminated with odour forming fungi. In this respect, many species of fungi are capable of O-methylation of the halophenols to produce the corresponding haloanisoles, the cause of 'mustiness' in many foods. To provide a comparison between the efficiencies of methylation (percentage conversion) by different species of the same genus these species will be discussed under the general introduction to the genus and not under individual species.

5.7.1 *Aspergillus*

Most foods, commodities and raw materials are subject to contamination by species of *Aspergillus*. The prevalence of this genus throughout nature is due to its tolerance of elevated temperatures and reduced a_w.[58] As a consequence, this genus is the major cause of food spoilage in tropical countries. About 150 species of *Aspergillus* are recognised. Species that produce 2,3,4,6-tetrachloroanisole in the presence of 2,3,4,6-tetrachlorophenol are *Aspergillus clavatus* (2 to 6%), *Aspergillus niger* (2 to 8%), *Aspergillus petrakii* (40 to 44%), *Aspergillus sydowii* (10 to 70%) and *Aspergillus versicolor* (6 to 80%).[59] Those species that produce 2,4,6-trichloroanisole from 2,4,6-trichlorophenol are *Aspergillus flavus* (35%), *Aspergillus sydowii* (19 to 77%) and *Aspergillus terreus* (1%).[60] These species had been isolated from broiler house litter that subsequently led to mustiness in broilers, and packaging material and floors of shipping containers that had resulted in mustiness in packaged and transported foods.[2]

Aspergillus candidus
Aspergillus candidus is occasionally associated with spoiled grain where it can produce the mycotoxin patulin.[58] It occurs most commonly on barley during malting and is also found on wheat and rice, pecan nuts, peanuts, copra and salted dried fish. The optimum temperature for growth is 25°C but this species can grow between 5 and 42°C. Its minimum a_w for growth is 0.88.[58]

Strains of *Asp. candidus* have been grown for 4 and 7 days on oats and wheat with a moisture content adjusted to 25%.[61] Seven compounds were identified including 2-methylfuran, 2-methylpropanol, 1-penten-3-ol, 2-methylbutanol, a dimethylbenzene, ethylbenzene and thujopsene. Unfortunately the effect of these compounds on the flavour of the grains was not discussed.

Aspergillus flavus
Aspergillus flavus is the main source of aflatoxins in susceptible foods[58] and reflecting its economic importance it is the most widely reported foodborne

fungus.[58] This fungus is abundant in the tropics where it has an affinity for nuts, oilseeds and cereal grains. *Asp. flavus* will grow at temperatures between 10 and 48°C and over the pH range of 2.1 to 11.2. Its minimum a_w for growth varies between 0.78 at 33°C and 0.84 at 25°C.[58]

Growth of *Asp. flavus* on wheat or oats led to the production of eight compounds on the wheat medium and nine on the oats. Six of these compounds were the same as those identified in cultures inoculated with *Asp. candidus*. The additional compounds were nitromethane, limonene and 1,3-octadiene.[61] However, when different strains of this fungus were grown on bread the only compounds detected were four compounds with intense musty odours.[55] As the presence of such musty odours would greatly affect the sensory quality of grain, an effort should be made to identify the compounds responsible. Of interest is *Aspergillus versicolor* which produced the same group of compounds when grown on wheat or oats.[61]

Aspergillus niger

Aspergillus niger is commonly associated with the post-harvest decay of fresh fruit where its black colonies are indicative of its presence. It is also frequently found on nuts, some grains, oilseeds and meat products. *Asp. niger* will grow over the temperature range of 6 to 47°C and at a pH of 2 at high a_w. This fungus is a xerophile and will germinate at a_w 0.77 at 35°C.[58] Inoculation of bread with *Asp. niger* led to the identification of seven compounds of which the major compounds were 1-octen-3-ol, 3-methylbutanol, nonanal and 3-octanone.[55] Two unidentified musty smelling compounds were also detected. Both compounds had been previously detected when *Asp. flavus* was grown on the same medium.

5.7.2 *Eurotium*

About 20 *Eurotium* species are known of which four, *Eurotium amstelodami*, *Eurotium chevalieri*, *Eurotium repens* and *Eurotium rubrum* are exceedingly common in all kinds of environments where just sufficient moisture exists to support fungal growth.[58] All four of these species had been, until the First International Workshop on *Penicillium* and *Aspergillus* held in the Netherlands in 1985, classified as members of the genus *Aspergillus*, and *Eur. rubrum* was known as *Aspergillus ruber*. All species of *Eurotium* are xerophilic.[58]

Five species of *Eurotium* have been found to O-methylate 2,3,4,6-tetrachlorophenol to 2,3,4,6-tetrachloroanisole. The species are *Eur. amstelodami* (2 to 39%), *Eur. chevalieri* (1 to 3%), *Eur. chevalieri* var. *intermedius* (45%), *Eur. repens* (27 to 74%) and *Eur. rubrum* (3 to 31%).[59] Those species that produce 2,4,6-trichloroanisole from 2,4,6-trichlorophenol are *Eur. chevalieri* (0.6%), *Eurotium herbariorum* (7%) and *Eur. repens* (12%).[60]

Eurotium amstelodami
Eur. amstelodami is a ubiquitous foodborne species that is usually associated with stored produce and is frequently found on cereals including wheat, rice, maize and corn snacks. The optimal temperature for growth of this fungus is 33 to 35°C and it has been reported to grow down to a_w 0.70 at 25°C.[58]

Eur. amstelodami has been associated with ketonic rancidity in desiccated coconut.[62] Compounds identified in the spoiled product were 2-hexanone, 2-heptanone, 2-octanone, 2-nonanone, 2-undecanone, 2-hexanol, 2-heptanol and 2-nonanol.[63] Sensory analyses indicated that 2-hexanone, 2-heptanone and 2-heptanol were the compounds responsible for the off-flavour. Growth of *Eur. amstelodami* on desiccated coconut led to the identification of all of the above compounds with the exception of 2-hexanol. In addition, 2-pentanone was also identified.[64] Analyses of the volatile fatty acids present in spoiled and unspoiled desiccated coconut showed that formation of the ketones corresponded to a decrease in the short chain acids in the spoiled product.[63] It was suggested that the first step in the metabolic process was the enzymic hydrolysis of the triacylglycerols to give the free fatty acids which were subsequently converted to the ketones by a modified β-oxidation. Reduction of the ketones by an alcohol dehydrogenase would yield the corresponding alcohols. *Eur. chevalieri* and *Eur. herbarium* were also isolated from the spoiled coconut and were shown to produce identical metabolites although some quantitative differences were apparent.[64]

5.7.3 *Penicillium*
Penicillium is the most diverse genus of fungi both in terms of numbers of species and range of habitats. It is also regarded as one of the major causes of food spoilage by microorganisms.[58] Most *Penicillium* species are ubiquitous and opportunistic, being able to grow in almost any environment and survive under a wide range of conditions of temperature, pH, a_w and redox potential.[58] Although the majority of *Penicillium* species are soil fungi, many species appear to have their primary natural habitat in cereal grains. A number of species are psychrotrophic and are capable of food spoilage at refrigeration temperatures.

Penicillium species are known to produce a range of volatile metabolites. In 1995, some 47 species grown on three different media were analysed for such compounds.[65] Unfortunately, the data obtained are far too extensive to be included, in total, in the current article. In addition, many of the species studied are not known to cause food spoilage. However, where a species is a known food contaminant, details of its volatile metabolites will be included.

Four species of *Penicillium* have been found to O-methylate 2,3,4,6-tetrachlorophenol to 2,3,4,6-tetrachloroanisole. These species are

Penicillium corylophilum (62 to 83%), *Penicillium crustosum* (2 to 16%), *Penicillium glabrum* (54%) and *Penicillium roqueforti* (7 to 65%).[59] By comparison, six species were found to convert 2,4,6-trichlorophenol to 2,4,6-trichloroanisole. The species were *Penicillium brevicompactum* (27%), *Penicillium citrinum* (28%), *Pen. corylophilum* (5%), *Pen. crustosum* (31%), *Penicillium janthinellum* (23%), and *Penicillium variabile* (6%).[60]

Penicillium camemberti, Penicillium caseicolum
Surface-ripened cheeses involving extensive mould growth during ripening such as brie and camembert occasionally develop earthy-musty off-flavours as they mature. *Penicillium camemberti* and *Penicillium caseicolum* are used in the production of such cheeses. Growth of commercial strains of these fungi on suitable media led to the identification of 13 compounds of which six had odours described as earthy, mushroom, geranium and potato-like.[66] Both species gave the same series of compounds and all but one of the compounds were present in both young and mature cultures. However, the majority of compounds were present in greater quantities in the mature cultures. Compounds with mushroom-like odours were 3-octanone, 1-octen-3-ol, those with geranium odours were 1,5-octadien-3-one and 1,5-octadien-3-ol, that with the earthy odour was 2-methylisoborneol and that with the potato-like odour was 2-methoxy-3-isopropylpyrazine. The last named compound was only found in aged cultures (4 to 6 weeks) where it was responsible for intense earthy/raw potato aromas in the cheese.[66] In other studies *Pen. camemberti* was shown to produce 10 compounds of which five had been previously identified.[65] The additional compounds were ethyl acetate, ethyl 3-methylpropanoate, 2-methylpropyl acetate, ethyl hexanoate and styrene. The presence of these compounds could produce an undesirable flavour and odour if present in some foods.

Penicillium digitatum, Penicillium expansum, Penicillium italicum
Penicillium digitatum and *Penicillium italicum* are two of the major causes of destructive rots in citrus fruits whereas *Penicillium expansum* is the principal cause of spoilage in some fruits.[58] Where these citrus fruits are used for processing, the presence of *Pen. digitatum* and *Pen. italicum* can lead to intense off-flavours in the products.[2] Such an off-flavour is the so called 'mouldy orange' flavour that is occasionally associated with orange juice. Similarly, stored apples and pears infected with *Pen. expansum* can develop intense, penetrating earthy odours as they decay.[67]

Pen. digitatum can grow between 6 and 37°C and has a minimum a_w for growth of 0.90 at 25°C.[58] Although volatile metabolites associated with citrus fruit spoiled by this fungi have not been identified, 12 compounds have been identified in cultures of *Pen. digitatum*. The compounds obtained from such cultures included the acetates of the C_2 to C_8 even-carbon alkanols together with isopropyl acetate, ethyl propanoate, ethyl 2-methylpropanoate, 2-methylpropyl acetate, 3-methylbutyl acetate and β-selinene.[65] However, it is unlikely that any of these compounds would account for the characteristic odour associated with citrus fruit spoiled by this fungus.

Pen. italicum can grow between −3 and 34°C and has a minimum a_w for germination of 0.86.[58] Again the volatiles associated with mouldy citrus have not been identified but 12 compounds have been found in cultures of this fungus.[65] The major metabolites formed were ethyl acetate, 3-methylbutanol and linalool together with 2-methylpropanol, 1-octene, ethyl butanoate, ethyl 2-methylbutanoate, 1-nonene, styrene, citronellene, β-farnesene and nerolidol.[65] The presence of some of these compounds, particularly linalool, could produce an unpleasant flavour in some contaminated food. However, their role in mouldy orange flavours is uncertain.

Pen. expansum can grow between −6 and 35°C and has a minimum a_w for germination of 0.82 to 0.83. This fungus has a very low requirement for oxygen. Although it is found principally on a wide variety of fruits, it is also found on a range of other foods including cheese and margarine.[58] Volatiles from such infected foods have not been identified; however, seven compounds have been obtained from cultures of this fungus.[65] The major metabolites formed were 2-methylpropanol, 3-methylbutanol and β-bergamotene together with β-pinene, 1-methoxy-3-methylbenzene zingiberene and β-bisabolene.[65] Geosmin has also been identified as a metabolite of this fungus.[67] In the author's laboratory, some samples of margarine contaminated with *Pen. expansum* had an earthy off-flavour but attempts to identify the compound responsible were unsuccessful.

Penicillium solitum

Penicillium solitum has the ability to grow at low temperatures and a_w but will not grow at 37°C. It is a recognised pathogen of pomaceous fruit but is found on a range of other foods including cheese and margarine. In some instances where this fungus has been found in low salt margarine its presence has been associated with a strong crushed ant-like odour and flavour.[68] Compounds found in the contaminated product were the odd chain aliphatic ketones 2-pentanone, 2-heptanone, 2-nonanone and 2-undecanone. 2-Heptanone, the major metabolite, was considered to be the principal source of the off-flavour. Interestingly, where this species was cultured on an agar and margarine medium, traces of 2-methylisoborneol and geosmin were identified.[68] The formation of the aliphatic ketones was considered to proceed by abortive β-oxidation of the C_6 to C_{12} fatty acids. Accordingly, it would appear that *Pen. solitum* is also strongly lipolytic.[68]

5.8 Future trends

Since the 1960s there has been a change in emphasis concerning the type of microorganisms involved in the formation of off-flavours in foods. In the

1960s, studies of the role of bacteria in the spoilage of meat, fish and dairy products were paramount. However, in the 1980s and 1990s there was a significant shift in interest towards problems of fungal origin. Actinomycetes had also been shown to play an important role in the cause of off-flavours in foods and beverages. Only yeasts appear to have been overlooked. Possibly in future years these organisms will receive greater attention. Their presence in foods could produce as yet unrecognised metabolites deleterious to the flavour quality of food.

Investigators have the skills and microbiological techniques to identify with certainty the organisms responsible for the production of odorous compounds in foods. In addition, analytical instrumentation is available that will permit the identification of minute quantities of off-flavour compounds in complex food systems. The excellent use of these skills and techniques has been extensively illustrated in this chapter. However, one field that continues to remain a challenge is in the metabolic pathways used by such organisms in the formation of off-flavour compounds in food. Researchers willing to undertake this challenge will have the opportunity to confirm or correct the many speculative pathways that exist in the literature. In addition they will have the opportunity to lift our understanding of the biochemistry of off-flavour development to the same level as that exhibited in the identification of the organisms and compounds responsible for such microbiologically derived off-flavours in foods.

5.9 References

1. SPRINGETT M B (1996), 'Formation of off-flavours due to microbiological and enzyme action', in Saxby M J (ed), *Food Taints and Off-flavours*, London, Blackie Academic and Professional, Chap. 9, pp 274–89.
2. WHITFIELD F B (1998), 'Microbiology of food taints', *Int J Food Sci Technol*, **33**, 31–51.
3. KRIEG N R (1984), *Bergey's Manual of Systematic Bacteriology*, Baltimore, Williams and Wilkins, Vol 1.
4. STUTZ H K, SILVERMAN G J, ANGELINI P and LEVIN R E (1991), 'Bacteria and volatile compounds associated with ground beef spoilage', *J Food Sci*, **56**, 1147–53.
5. MILLER A, SCANLAN R A, LEE J S and LIBBEY L M (1973), 'Volatile compounds produced in sterile fish muscle (*Sebastes melanops*) produced by *Pseudomonas putrefaciens*, *Pseudomonas fluorescens*, and an *Achromobacter* species', *Appl Microbiol*, **26**, 18–21.
6. JENSEN N (2000), '*Alicyclobacillus* in Australia', *Food Austral*, **52**, 282–5.
7. PETTIPHER G L, OSMUNDSON M E and MURPHY J M (1997), 'Methods for the detection and enumeration of *Alicyclobacillus acidoterrestris* and investigation of growth and production of taint in fruit juice and fruit juice-containing drinks', *Lett Appl Microbiol*, **24**, 185–9.
8. BORLINGHAUS A and ENGEL R (1997), '*Alicyclobacillus* incidence in commercial apple juice concentrate (AJC) supplies – method development and validation', *Fruit Process*, **7**, 262–6.

9. JENSEN N and WHITFIELD F B (2003), 'Role of *Alicyclobacillus acidoterrestris* in the development of a disinfectant taint in shelf stable fruit juice', *Lett Appl Microbiol*, **36**, 9–14.
10. HUANG Z, DOSTAL L and ROSAZZA J P N (1993), 'Mechanisms of ferulic acid conversions to vanillic acid and guaiacol', *J Biol Chem*, **268**, 23954–8.
11. EDWARDS J and KOSIKOWSKI F V (1983), 'Bitter compounds from Cheddar cheese', *J Dairy Sci*, **66**, 727–34.
12. SKEAN J D and OVERCAST W W (1960), 'Changes in the paper electrophoretic protein patterns of refrigerated skim milk accompanying growth of three *Pseudomonas* species', *Appl Microbiol*, **8**, 335–8.
13. FREEMAN L R, SILVERMAN G J, ANGELINI P, MERRITT C and ESSELEN W B (1976), 'Volatiles produced by microorganisms isolated from refrigerated chicken at spoilage', *Appl Environ Microbiol*, **32**, 222–31.
14. PITTARD B T, FREEMAN L R, LATER D W and LEE M L (1982), 'Identification of volatile organic compounds produced by fluorescent Pseudomonads on chicken breast muscle', *Appl Environ Microbiol*, **43**, 1504–6.
15. VIEHWEG S H, SCHMITT R E and SCHMITT-LORENZ W (1989), 'Microbial spoilage of refrigerated fresh broilers. Part VII. Production of off-odours from poultry skin by bacterial isolates', *Lebensm-Wiss u-Technol*, **22**, 356–67.
16. JEON I J, 'Undesirable Flavors in Dairy Products' (1996), in Saxby M J (ed), *Food Taints and Off-flavours*, London, Blackie Academic and Professional, Chap 5, pp 122–49.
17. KWAK H S, JEON I J and PERNG S K (1989), 'Statistical patterns of lipase activities on the release of short-chain fatty acids in Cheddar cheese slurries', *J Food Sci*, **54**, 1559–64.
18. PEREIRA J N and MORGAN M E (1958), 'Identity of esters produced in milk cultures by *Pseudomonas fragi*', *J Dairy Sci*, **42**, 1201–5.
19. REDDY M C, BILLS D D, LINDSAY R C, LIBBEY L M, MILLER A and MORGAN M E (1968), 'Ester production by *Pseudomonas fragi*. I. Identification and quantification of some esters produced in milk cultures', *J Dairy Sci*, **51**, 656–9.
20. MENCHER J R and ALFORD J A (1967), 'Purification and characterization of the lipase of *Pseudomonas fragi*', *J Gen Microbiol*, **48**, 317–28.
21. PITAS R E, SAMPUGNA J and JENSEN R G (1967), 'Triglyceride structure of cow milk fat. I. Preliminary observations on the fatty acid composition of positions 1, 2 and 3', *J Dairy Sci*, **50**, 1332–6.
22. HOSONO A and ELLIOTT J A (1974), 'Properties of crude ethyl ester-forming enzyme preparations from some lactic acid and psychrotrophic bacteria', *J Dairy Sci*, **57**, 1432–7.
23. HOSONO A, ELLIOTT J A and MCGUGAN W A (1974), 'Production of ethyl esters by some lactic acid and psychrotrophic bacteria', *J Dairy Sci*, **57**, 535–9.
24. CASTELL C H, GREENOUGH M F and DALE J (1959), 'The action of *Pseudomonas* on fish muscle: 3. Identification of organisms producing fruity and oniony odours', *J Fisheries Res Board Canada*, **16**, 13–9.
25. MILLER A, SCANLAN R A, LEE J S and LIBBEY L M (1973), 'Identification of the volatile compounds produced in sterile fish muscle (*Sebastes melanops*) by *Pseudomonas fragi*', *Appl Microbiol*, **25**, 952–5.
26. DAINTY R H, EDWARDS R A, HIBBARD C M and MARNEWICK J J (1989), 'Volatile compounds associated with microbial growth on normal and high pH beef stored at chill temperatures', *J Appl Bacteriol*, **66**, 281–9.
27. TURNER A W (1927), '*Achromobacter perolens* (n.sp.) the cause of 'mustiness' in eggs', *Austral J Exp Biol Med Sci*, **4**, 57–60.
28. SZYBALSKI W (1950), 'A comparative study of bacteria causing mustiness in eggs', *Nature (London)*, **165**, 733–4.

29. CASTELL C H, GREENOUGH M F and JENKIN N L (1957), 'The action of *Pseudomonas* on fish muscle: 2. Musty and potato-like odours', *J Fisheries Res Board Canada*, **14**, 775–82.
30. MILLER A, SCANLAN R A, LEE J S, LIBBEY L M and MORGAN M E (1973), 'Volatile compounds produced in sterile fish muscle (*Sebastes melanops*) by *Pseudomonas perolens*', *Appl Microbiol*, **25**, 257–61.
31. CHENG T-B, REINECCIUS G A, BJORKLUND J A and LEETE E (1991), 'Biosynthesis of 2-methoxy-3-isopropylpyrazine by *Pseudomonas perolens*', *J Agric Food Chem*, **39**, 1009–12.
32. WHITFIELD F B, JENSEN N and SHAW K J (2000), 'Role of *Yersinia intermedia* and *Pseudomonas putida* in the development of a fruity off-flavour in pasteurized milk', *J Dairy Res*, **67**, 561–9.
33. LEVINE M and ANDERSON D Q (1932), 'Two new species of bacteria causing mustiness in eggs', *J Bacteriol*, **23** 337–47.
34. TOMPKIN R B and SHAPARIS A B (1972), 'Potato aroma of lamb carcasses', *Appl Microbiol*, **24**, 1003–4.
35. MORGAN M E, LIBBEY L M and SCANLAN R A (1972), 'Identity of the musty-potato aroma compound in milk cultures of *Pseudomonas taetrolens*, *J Dairy Sci*, **55**, 666.
36. GALLOIS A, KERGOMARD A and ADDA J (1988), 'Study of the biosynthesis of 3-isopropyl-2-methoxypyrazine produced by *Pseudomonas taetrolens*', *Food Chem*, **28**, 299–309.
37. SHEWAN J M (1962), 'The bacteriology of fresh and spoiling fish and some related chemical changes', in Hawthorn J and Muil Leitch J (eds), *Recent Advances in Food Science*, London, Butterworths, Vol.1, pp 167–93.
38. SHAW B G and SHEWAN J M (1968), 'Psychrophilic spoilage bacteria of fish', *J Appl Bacteriol*, **31**, 89–96.
39. GRAM L, TROLLE G and HUSS H H (1987), 'Detection of specific spoilage bacteria from fish stored at low (0°C) and high (20°C) temperatures', *Int J Food Microbiol*, **4**, 65–72.
40. SNEATH P H A (1986), *Bergey's Manual of Systematic Bacteriology*, Baltimore, Williams and Wilkins, Vol 2.
41. SCHAFTER M L, PEELER J T, BRADSHAW J G, HAMILTON C H and CARVER R B (1985), 'Gas chromographic detection of D-(−)-2,3-butanediol and butyric acid produced by sporeformers in cream-style corn and canned beef noodle soup – Collatorative study', *J Assoc Off Anal Chem*, **68**, 626–31.
42. KINDERLERER J L and KELLARD B (1987), 'Alkylpyrazines produced by bacterial spoilage of heat-treated and gamma-irradiated coconut', *Chem Ind*, pp 567–8.
43. KENNEDY M B and LENNARZ (1979), 'Characterization of the extracellular lipase of *Bacillus subtilus* and its relationship to a membrane-bound lipase found in a mutant strain', *J Biol Chem*, **254**, 1080–9.
44. KOSUGE T and KAMIYA H (1962), 'Discovery of a pyrazine in a natural product: Tetramethylpyrazine from cultures of a strain of *Bacillus subtilis*', *Nature (London)*, **193**, 776.
45. DAINTY R H and HIBBARD C M (1980), 'Aerobic metabolism of *Brochothrix thermosphacta* growing on meat surfaces and in laboratory media', *J Appl Bacteriol*, **48**, 387–96.
46. DAINTY R H and HIBBARD C M (1983), 'Precursors of the major end products of aerobic metabolism of *Brochothrix thermosphacta*', *J Appl Microbiol*, **55**, 127–33.
47. GALLOIS A and GRIMONT P A D (1985), 'Pyrazines responsible for the potato like odor produced by some *Serratia* and *Cedecea* strains', *Appl Environ Microbiol*, **50**, 1048–51.

48. JENSEN N, VARELIS P and WHITFIELD F B (2001), 'Formation of guaiacol in chocolate milk by the psychrotrophic bacterium *Rahnella aquatilis*', *Lett Appl Microbiol*, **33**, 339–43.
49. DAINTY R H, EDWARDS R A and HIBBARD C M (1989), 'Spoilage of vacuum-packed beef by a *Clostridium* sp', *J Sci Food Agric*, **49**, 473–86.
50. WHITFIELD F B, LAST J H and TINDALE C R (1982), 'Skatole, indole and *p*-cresol: components in off-flavoured frozen French fries', *Chem Ind*, pp 662–3.
51. FELLERS C R and CLOUGH R W (1925), 'Indol and skatol determination in bacterial cultures', *J Bacteriol*, **10**, 105–33.
52. ELSDEN S R, HILTON M G and WALLER J M (1976), 'The end products of the metabolism of aromatic amino acids by Clostridia', *Arch Microbiol*, **107**, 283–8.
53. JENSEN B B and JENSEN M T (1995), 'Effect of diet composition on microbial production of skatole in the hindgut of pigs and its relation to skatole in backfat', in Nunes A F, Portugal A V, Costa J P and Ribeiro J R (eds), *Protein Metabolism and Nutrition*, Santarem, Estacao Zootecnica National, pp 489–94.
54. GERBER N N (1979), 'Volatile substances from *Actinomycetes*: their role in the odor pollution of water', *CRC Crit Rev Microbiol*, **7**, 191–214.
55. HARRIS N D, KARAHADIAN C and LINDSAY R C (1986), 'Musty aroma compounds produced by selected molds and *Actinomycetes* on agar and whole wheat bread', *J Food Protection*, **49**, 964–70.
56. WASOWICA E, KAMINSKI E, KOLLMANNSBERGER H, NITZ S, BERGER R G and DRAWERT F (1988), 'Volatile components of sound and musty wheat grains', *Chem Mikrobiol Technol Lebensm*, **11**, 161–8.
57. WHITFIELD F B, SHAW K J, GIBSON A M and MUGFORD D C (1991), 'An earthy off-flavour in wheat flour: geosmin produced by *Streptomyces griseus*', *Chem Ind*, pp 841–2.
58. PITT J I and HOCKING A D (1996), *Fungi and Food Spoilage*, Melbourne, Blackie Academic and Professional.
59. GEE J M and PEEL J L (1974), 'Metabolism of 2,3,4,6-tetrachlorophenol by microorganisms from broiler house litter', *J Gen Microbiol*, **85**, 237–43.
60. TINDALE C R, WHITFIELD F B, LEVINGSTON S D and NGUYEN T H L (1989), 'Fungi isolated from packaging materials: their role in the production of 2,4,6-trichloroanisole', *J Sci Food Agric*, **49**, 437–47.
61. BÖRJESSON T, STÖLLMAN U and SCHNÜRER J (1992), 'Volatile metabolites produced by six fungal species compared with other indicators of fungal growth on cereal grains', *Appl Environ Microbiol*, **58**, 2599–602.
62. KINDERLERER J L (1984), 'Spoilage in desiccated coconut resulting from growth of xerophilic fungi', *Food Microbiol*, **1**, 23–8.
63. KELLARD B, BUSFIELD D M and KINDERLERER J L (1985), 'Volatile off-flavour compounds in desiccated coconut', *J Sci Food Agric* **36**, 415–20.
64. KINDERLERER J L and KELLARD B (1984), 'Ketonic rancidity in coconut due to xerophilic fungi', *Phytochem*, **12**, 2847–9.
65. LARSEN T O and FRISVAD J C (1995), 'Characterization of volatile metabolites from 47 *Penicillium* taxa', *Mycol Res* **99**, 1153–66.
66. KARAHADIAN C, JOSEPHSON D B and LINDSAY R C (1985), 'Volatile compounds from *Penicillium* sp. contributing musty-earthy notes to Brie and Camembert cheese flavors', *J Agric Food Chem*, **33**, 339–43.
67. MATTHEIS J P and ROBERTS R G (1992), 'Identification of geosmin as a volatile metabolite of *Penicillium expansum*', *Appl Environ Microbiol*, **58**, 3170–2.
68. HOCKING A D, SHAW K J, CHARLEY N J and WHITFIELD F B (1998), 'Identification of an off-flavour produced by *Penicillium solitum* in margarine', *J Food Mycol*, **1**, 23–30.

6
Oxidative rancidity as a source of off-flavours

R. J. Hamilton, formerly of Liverpool John Moores University, UK

6.1 Introduction

The major components of food are carbohydrates, proteins and lipids. Off-flavours can be produced from the carbohydrate and the protein part of the food, but it is mainly the lipid portion of the food which gives rise to off-flavours. Rancidity is sometimes defined as the subjective organoleptic appraisal of off-flavours in foods. It is subjective because the ability to perceive an off-flavour varies from person to person. The flavour threshold of a compound can vary greatly so that it is possible for a small amount of an off-flavour to give an unpleasant taste to a large quantity of a foodstuff. The flavour threshold is defined as the minimum quantity of a substance, which can be detected by 50% of the taste panel (Hamilton, 2001).

This chapter will describe the three main chemical mechanisms involved in the oxidation of lipid molecules. Irrespective of the oxidation mechanism, it is recognised that the lipid molecule is oxidised to an intermediate which has no odour or flavour in itself. However, it is the breakdown of this lipid intermediate as in Eqn [6.1] which gives the off-flavour:

$$\text{Lipid} \to \text{Intermediate} \to \text{Secondary reaction products.} \qquad [6.1]$$

Unsaturated lipids can be oxidised in the dark and at room temperature. These conditions are associated with the production of free radicals and the mechanism is known as autoxidation (see Section 6.3). Depending on the type of food, the unsaturated nature of the lipid portion and the amounts of antioxidants present in the food, these reactions exhibit a lag phase (induction period). If some measure of the oxidation, for example peroxide value or oxygen uptake, is measured with time, a graph like that shown

Oxidative rancidity as a source of off-flavours   141

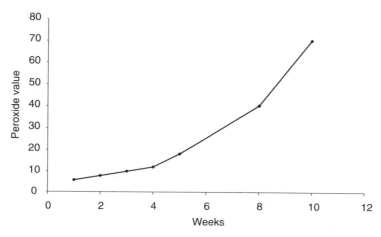

Fig. 6.1 Induction curve for the room temperature oxidation of Chilean fish oil.

in Fig. 6.1. can be produced which is sometimes called the induction curve (Simpson, 1999). It can be seen that the peroxide value remains low for several hours or days before it increases rapidly. The time to reach this point in the graph is known as the induction period. In the instance shown the fish oil does not show a dramatic change from lag phase to the rapidly changing peroxide value. This change is more pronounced in animal fats such as lard. The foodstuff will give a distinct taste or odour when the peroxide value rises sharply but expert tasters will be able to sense the first off-flavours earlier in the lag phase (Hamilton et al., 1998).

Unsaturated lipids also undergo oxidation in the light especially when certain molecules, known as sensitisers, are present. Here there is little or no induction period. The two mechanisms in light conditions are known as Type I and Type II photo-oxidations (see Section 6.4). The intermediates formed in Type II photo-oxidations are different from the intermediates in autoxidation with the result that the off-flavours which come from the degradation products in Eqn [6.1] are different. Finally, the third route to the intermediate that involves the enzyme lipoxygenase is dealt with in Section 6.5.

Rancidity can be associated with the effect of water on the lipid constituents of food. Since water and fat can sit together for a long time without any hydrolysis occurring, it is the presence of a catalyst which leads to off-flavours. One type of catalyst is the enzyme lipase, which is usually introduced into the foodstuff by microbiological contamination. The off-flavours produced by microorganisms are dealt with in Chapter 5. It should be stressed that most rancid off-flavours come from the oxidation of the lipid molecules in food.

6.2 Oxidation

Almost all organic molecules undergo oxidation when brought into contact with oxygen. Lipid molecules in clay cooking pots from archaeological digs have been shown to be derived from the cabbages of the foodstuffs cooked by the iron-age dwellers in those sites. That such molecules can remain intact without being converted into carbon dioxide and water shows how stable some lipids are (Dudd, 1999). However, the drive towards 'healthy' foods has resulted in the incorporation of more unsaturated fatty acids in the diet. These newer foods are likely to be oxidised more easily and have to be better protected against oxidation.

Oxidation can occur by chemical, photochemical and enzyme reactions. Off-flavours are usually derived from unsaturated fatty acids either free or in bound form as esters in acylglycerols and as amides in ceramides and sphingolipids (Figs 6.2 and 6.3) (Gunstone, 1996). It is generally accepted

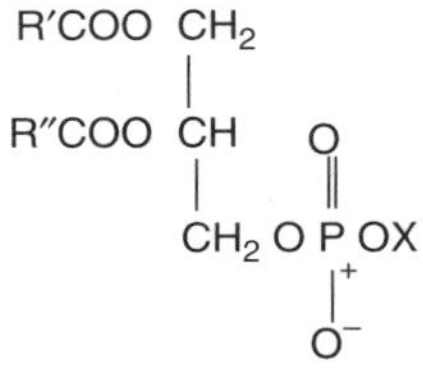

$$R'COO-CH_2$$
$$R''COO-CH$$
$$CH_2-O-P(=O)(O^-)-OX$$

COMPOUND

where
X = H — Phosphatidic acid

X = $CH_2CH(NH_2)COOH$ — Phosphatidyl serine

X = $CH_2CH_2NH_2$ — Phosphatidyl ethanolamine (cephalin)

X = $CH_2CH_2\overset{+}{N}(CH_3)_3$ — Phosphatidyl choline (lecithin)

Fig. 6.2 Structures of acylglycerols.

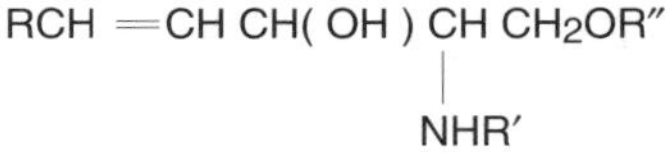

$$RCH=CH-CH(OH)-CH-CH_2OR''$$
$$\quad\quad\quad\quad\quad\quad\quad\quad\quad |$$
$$\quad\quad\quad\quad\quad\quad\quad\quad NHR'$$

where

R' = H   R'' = H   sphingenine

R' = COR   R'' = H   ceramides

R' = COR'   R'' = glucose or galactose   cerebrosides

Fig. 6.3 Structures of sphingolipids.

Table 6.1 Fat content of foods (adapted from Crawford, 1985)

	Fat content (g per 100 g)	Saturated fatty acid content (g per 100 g)
Corn oil	99.9	16.4
Butter	82.0	49.0
Margarine (sunflower)	81.0	10.1
Oil spread (olive)	59.0	14.0
Peanuts	49.0	9.2
Double cream	48.2	28.8
Whole milk	3.8	2.3
Cheddar cheese	33.5	20.0
Potato crisps (chips)	34.0	16.0
Grilled streaky bacon	36.0	14.5
Plain biscuit	16.6	7.9
Chocolate biscuit	27.6	16.6
Uncooked smoked mackerel	16.3	3.1
Uncooked cod	0.7	0.1

that the reaction of atmospheric oxygen is the principal cause of fat deterioration. Many foods have a high lipid content, for example butter or margarine (Table 6.1), whilst other foods such as dried milk, milk products, crisps and baked goods have a significant lipid content. Such foods are especially prone to becoming rancid during storage

6.3 Autoxidation

The mechanism of autoxidation has three stages:

- initiation
- propagation
- termination

At one time it was believed that the lipid molecule represented as LH could react directly with oxygen as in Eqn [6.2]:

$$LH + {}^3O_2 \rightarrow LOOH \qquad [6.2]$$

where $^3O_2$ represents triplet oxygen.

It is now known that such an interaction between triplet oxygen and a lipid does NOT happen. The initiation stage results in the formation of two free radicals as in Eqn [6.3]:

$$LH \rightarrow L\cdot + H\cdot \qquad [6.3]$$

where L· and H· are free radicals.

The formation of this first free radical in the chain L· may occur by several different reactions. One possibility is that a metal atom can catalyse this first reaction, e.g.

$$M^{3+} + LH \rightarrow M^{2+} + H^+ + L\cdot \qquad [6.4]$$

The metal ion can come from the packaging or from processing conditions of the food or it can be part of the haem complex. Once the free radical (L·) is formed, propagation steps lead to the first isolable product, viz. the hydroperoxide. This is the Intermediate quoted in Eqn [6.1]:

$$L\cdot + O_2 \rightarrow LO_2\cdot \qquad [6.5]$$

$$LO_2\cdot + L''H \rightarrow LO_2H + L''\cdot \qquad [6.6]$$

These equations show that the peroxy free radical ($LO_2\cdot$) can attack the original lipid molecule LH or another lipid molecule L''H to break the LH bond with the formation of further alkyl free radicals L· or L''·.

Once the hydroperoxide is formed, it can break down in a number of ways. Equations [6.7], [6.8] and [6.9] show how a large number of free radicals are generated and in so doing help to form a chain reaction:

$$LOOH \rightarrow LOO\cdot + H\cdot \qquad [6.7]$$

$$LOOH \rightarrow LO\cdot + OH\cdot \qquad [6.8]$$

$$2\,LOOH \rightarrow LO\cdot + H_2O + LOO\cdot \qquad [6.9]$$

These steps generate such a large flux of free radicals that they create the situation where autoxidation can run on until considerable off-flavours have been produced.

The alternative ways of producing free radicals, for example by photo-oxidation or by the reaction of the lipoxygenase enzymes, can also feed into the autoxidation processes. The classical free radical reactions involved in polymerisation reactions for the formation of industrial polymers always have steps which result in the chain being broken. These termination reactions also occur in autoxidation but are normally important where the oxygen concentration is low:

$$LOO\cdot + LOO\cdot \rightarrow \text{stable molecules or relatively stable free radicals} \qquad [6.10]$$

$$LOO\cdot + L\cdot \rightarrow \text{stable molecules or relatively stable free radicals} \qquad [6.11]$$

$$L\cdot + L\cdot \rightarrow \text{stable molecules or relatively stable free radicals} \qquad [6.12]$$

Antioxidants have a much more effective role in causing termination of autoxidation because they react to give a free radical which is much more stable than the alkyl, peroxy or hydroperoxy free radical as shown in Eqn [6.13]:

$$L\cdot + AH \rightarrow RH + A\cdot \qquad [6.13]$$

where AH represents the antioxidant molecule and A· is a much more stable free radical derived from AH.

In this case, this stable free radical A· does not enter into the propagation reactions [3.8] and [3.9]. The hydroperoxide, which results from autoxidation, does not have an off-flavour or taste. Indeed the hydroperoxides, even those from photo-oxidation or from lipoxygenase, do not have a taste but they break down readily to hydrocarbons, alcohols, ketones and aldehydes as shown in Fig. 6.4.

The hydroperoxide breaks down to form an alkoxy free radical. From this radical, it is possible to show mechanisms by which the fatty acid chain is broken down with the production of an aldehyde like hexanal, an alcohol like octenol or a ketone. Some of the typical short chain compounds which are liberated by these reactions are given in Table 6.2. It should be noted that, at the same time as these volatile compounds are formed, there remains a more polar component which comes from the carboxylic end of the fatty acid. These compounds therefore have a carboxylic acid group at one end and either an alcohol, a carboxylic acid or an aldehyde group at the other end. Whilst not flavour volatiles, they may be a contributory factor in the production of a taint.

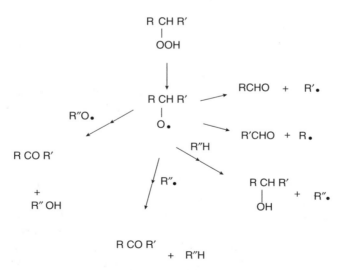

Fig. 6.4 Secondary oxidation products from the decomposition of hydroperoxides.

Table 6.2 Volatiles from the autoxidation of lipids

Compound	Source	Reference
γ-Decalactone	Peach and apricot	Buttery, 1989
Decanal	Whole milk	Buttery, 1989
Deca-2t,4c,7c-trienal	Fish oil	Fujimoto, 1989
Hepta-2,4-dienal	Butter oil	Frankel, 1998
Hex-2-enal	Apple	Buttery, 1989
Hex-3-enal	Stored boiled fish	Buttery, 1989
Nona-3,6-dienal	Stored boiled trout	Milo and Grosch, 1993
Non-2-enal	Cooked carrot	Buttery, 1989
Non-1-en-4-one	Bell pepper	Buttery, 1989
Octa-1,5-dien-3-ol	Fish tissues	Frankel, 1998
Oct-1-en-3-one	Mushroom	Buttery, 1989

One of the most damaging situations for the breakdown of the hydroperoxide is produced when the reaction is catalysed by metal ions as shown in Eqn [6.14]:

$$M^{2+} + LOOH \rightarrow LO\cdot + OH^- + M^{3+} \qquad [6.14]$$

The metals copper and iron are the most significant in the breakdown reactions. In compounding the disadvantage, the M^{2+} ion is oxidised to M^{3+}, which is also capable of breaking down the hydroperoxide as shown in Eqn [6.15]

$$M^{3+} + LOOH \rightarrow LOO\cdot + H\cdot + M^{2+} \qquad [6.15]$$

So a vicious cycle is set up which will continue to operate until all the hydroperoxide is broken down. It is estimated that 0.05 ppm of copper will reduce the keeping time of lard by 50% at a temperature of 98°C. Hydroperoxides are good oxidising agents and can cause the breakdown of fat-soluble vitamins A, D and E. They can also oxidise the sulfur side groups in proteins with the formation of S–S bridges between protein molecules, which results in the loss of nutritional quality and a tougher texture (Hole, 2001).

6.4 Photo-oxidation

Oxidation of lipids can occur in the light when certain compounds known as sensitisers, for example phaeophytins, myoglobin, riboflavin and erythrosine, are present. These compounds function by converting triplet oxygen to singlet oxygen. Singlet oxygen can react directly with the lipid molecule, in contrast to the situation in autoxidation where triplet oxygen will not react directly with the lipid molecule. In Type 1 photo-oxidation,

Oxidative rancidity as a source of off-flavours   147

the same types of hydroperoxide are formed as in autoxidation. Riboflavin is one of the photosensitisers which is implicated in Type I oxidations. In addition it should be noted that primary and secondary antioxidants do not have the effect that they have in autoxidation of slowing down the reaction. In Type 2 photo-oxidation, an ene addition reaction occurs where the oxygen adds across the double bond. Type II sensitisers have the effect shown in Eqn [6.16] that they convert triplet oxygen to singlet oxygen which can then react directly with the fatty acid molecule (Eqn [6.17]):

$$^3\text{Sens}^* + {}^3\text{O}_2 \rightarrow {}^1\text{Sens} + {}^1\text{O}_2^* \qquad [6.16]$$

$$^1\text{O}_2^* + \text{LH} \rightarrow \text{LOOH} \qquad [6.17]$$

This has the effect of producing hydroperoxides (Fig. 6.5) which are different from those formed during autoxidation. This in turn leads to different volatile breakdown products and to different taints and off-flavours. Type 2 photo-oxidation is faster than autoxidation with relative rates of oxidation for the common unsaturated acids shown in Tables 6.3 and 6.4.

Photo-oxidation is also different from autoxidation in that there is no lag (induction) period. The hydroperoxides build up quickly as soon as

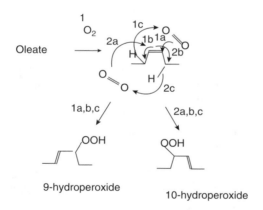

Fig. 6.5 Photo-oxidation mechanism (Type II).

Table 6.3 Relative reactivity of fatty acids with singlet oxygen

Lipid	Singlet oxygen reactivity
Oleic acid	1.1
Linoleic acid	1.9
Linolenic acid	2.9
Arachidonic acid	3.5

Table 6.4 Relative rates of autoxidation

Lipid	Relative rate
Oleate ester	1
Linoleate ester	41
Linolenate ester	98
Arachidonate ester	195

Table 6.5 Dissolved oxygen content of marine oil (adapted from MacFarlane, 2001)

Time (min)	Oxygen content (% oil in dark)	Oxygen content (% oil in light)
0	22	22
18	22	20
36	21	12
60	19	6
90	19	3
120	19	1

oxidation starts. This can result in a very rapid increase in peroxide value as shown by MacFarlane in Table 6.5. He followed the oxidation of a marine oil at room temperature by measuring the oxygen content of the oil. He started by saturating the oil with oxygen. Table 6.4 shows how the photo-oxidation leads to a rapid loss of dissolved oxygen when the oil is in the light. Amongst the taints associated with photo-oxidation is 'skunkiness' in beer. Hydrolysis and oxidation of isohumulones found in beer gives rise to 2,7-dimethyl-2,6-octadiene and 4-methyl-2-pentenal.

A 'cardboard-like' taste is described in milk left out in the sun (Allen, 1994). This latter taste is also referred to by other authors as 'burnt' or 'burnt feathers'. Riboflavin in the milk can be converted to a triplet form by light and then photo-oxidation of the fatty acids and methionine takes place. Carbonyl compounds from the oxidation of fatty acids and methional from the oxidation of the methionine can interact to give the off-flavour. Tryptophan is also known to be oxidised to o-aminoacetophenone, which is postulated to be the cause of the off-flavour in sterile milk. When butter is photo-oxidised, 3-methylnonane-2,4-dione can be isolated and the off-flavour attributed to this class of compounds. Frankel (1998) claims that furanoid fatty acids are the precursors of the diones which are found in stored boiled beef. Sulfur-containing compounds are often photo-oxidised in oils and fats and give rise to 'rubbery' off-flavours.

Fan, Tang and Wohlmann (1983) have shown that photo-oxidation with chlorophyll as a photosensitiser can be postulated as the cause of off-flavour of potato crisps. They suggested that the active component was dec-1-yne which they believe came from the oxidation of sterculic acid (a cyclopropane ring containing fatty acid). Normally carotenoids are viewed as being capable of controlling photo-oxidation reactions, because they react faster with singlet oxygen than the unsaturated lipid. This also means that the oils will be bleached and by the time the chlorophyll is used up the fatty acid becomes unprotected and will undergo oxidation itself in its turn. However, Min and Lee (1988) reported that tocopherols can minimise the oxidation of olive oil containing chlorophyll if the olive oil is left in the light.

6.5 Lipoxygenase (LOX)

No one reaction has been recorded as the sole initiation mechanism for autoxidation. Any reaction which produces free radicals could ultimately feed free radicals into the autoxidation pathway as has been intimated to occur in photo-oxidation. A similar claim can be made for the lipoxygenase route. Lipoxygenases are found widely in plants and animals and one of the first to be purified was soyabean seed lipoxygenase (Grosch, 1982). This enzyme catalyses the attack of oxygen on the pentadiene double bond system found in linoleic acid. In contrast to the autoxidation route, the enzyme causes the reaction to be both regiospecific and stereospecific (see Fig. 6.6; Gardiner, 1989). When one partially purified soyabean lipoxygenase was allowed to catalyse the reaction with linoleic acid, the regiospecificity shows itself to be 96% of the 13-hydroperoxide and 4% of the 9-positional isomer (see Table 6.6). Maize, wheat and pea lipoxygenases give different proportions of these two isomers.

Lipoxygenase activity requires the presence of free polyunsaturated fatty acids. Linoleic acid is the most common substrate in plant foods. The enzyme occurs in a variety of isozymes, which often vary in optimum pH, as well as product and substrate specificity. Four isozymes have been isolated from soybeans. Soy isozyme 1 has an optimum pH of 9.0. It only acts on free polyunsaturated fatty acids and it forms 9- and 13-hydroperoxides in the ratio of 1:9 at room temperature. Soy isozyme 2 has an optimum pH of 6.8, it acts on triglycerides as well as free polyunsaturated fatty acids and it forms 9- and 13-hydroperoxide in the ratio of about 1:1 at room temperature. Soy isozyme 3 is similar to isozyme 2, but its activity is inhibited by calcium ions, whereas lipoxygenase 2 is stimulated by the metal. Lipoxygenase 4 is very similar to isozyme 3, but can be separated by gel chromatography or electrophoresis. Lipoxygenase isozymes are commonly classified as type 1, which have an optimum pH in the alkaline region and are specific for free fatty acids, and type 2, which have optimum activity at neutral pH and cause co-oxidation of carotenoids. The ability of lipoxygenase type 2 to bleach carotenoids has found practical application in the

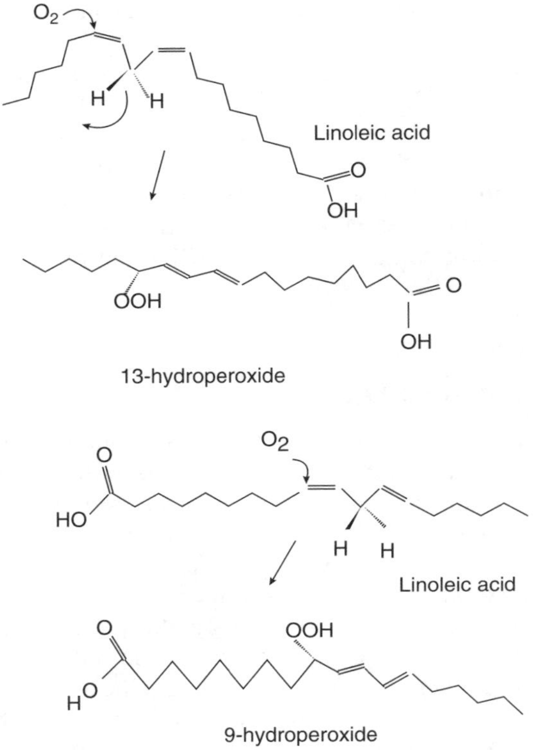

150   Taints and off-flavours in food

Fig. 6.6   Oxidation of linoleic acid by lipoxygenase.

Table 6.6   Regiospecificity and stereospecificity of lipoxygenase enzymes (adapted from: (x) Van Os *et al.*, 1979a; (y) Van Os *et al.*, 1979b; (z) Kuhn *et al.*, 1971)

	Regiospecific ratio of isomers		Stereospecific ratio of isomers			
	13 OOH	9 OOH	13 S OOH	13 R OOH	9 S OOH	9 R OOH
Soya LOX 1 (x)	97.5	2.5	97	3	61	39
Soya LOX 2 (y)	35	65	87	13	22	78
Maize LOX	14	86	79	21	94	6
Wheat LOX (z)	15	85	66	33	97.5	2.5

addition of soya flour to wheat flour in order to bleach the flour in the manufacture of white bread.

In plant tissues, various enzymes occur that cause the conversion of hydroperoxides to other products, some of which are important as flavour compounds. These enzymes include hydroperoxide lyase which catalyses the formation of aldehydes and oxo acids, hydroperoxide-dependent peroxygenase and epoxygenase, which catalyse the formation of epoxy and hydroxy fatty acids, and hydroperoxide isomerase, which catalyses the formation of epoxyhydroxy fatty acids and trihydroxy fatty acids. Lipoxygenase produces similar flavour volatiles to those produced during autoxidation, although the relative proportions of the products may vary widely depending on the specificity of the enzyme conditions.

Lipoxygenase molecules contain one atom of iron. The iron atom is in the high spin Fe(II) state in the native resting form of lipoxygenase, and it must be oxidised to Fe(III) by the reaction product, fatty acid hydroperoxides or hydrogen peroxide before it is active as an oxidation catalyst. As a consequence of this requirement for oxidation of the iron in the enzyme, a lag period is observed, when the enzyme is used with pure fatty acid substrates. The active enzyme abstracts a hydrogen atom stereospecifically from the intervening methylene group of a polyunsaturated fatty acid in a rate limiting step, with the iron being reduced to Fe(II). The enzyme–alkyl radical complex is then oxidised by molecular oxygen to an enzyme–peroxy radical complex under aerobic conditions, before electron transfer from the ferrous atom to the peroxy group occurs. Protonation and dissociation from the enzyme allow the formation of the hydroperoxide. Under anaerobic conditions, the alkyl radical dissociates from the enzyme–alkyl radical complex, and a mixture of products including dimers, ketones and epoxides is produced by radical reactions.

6.6 Ketonic rancidity and metal-catalysed lipid oxidation

Ketonic rancidity is a problem that can be encountered with some products such as desiccated coconut which contain short-chain saturated fatty acids. Moulds such as *Eurotium amstelodami* degrade triglycerides in the presence of limited amounts of air and water. Free fatty acids are liberated initially and these subsequently suffer β-oxidation with the formation of methyl ketones and aliphatic alcohols. A musty, stale note in the product is characteristic of ketonic rancidity (Kellard et al., 1985).

All materials of biological origin contain small amounts of transition metals, which cannot be completely removed by normal food processing. Transition metals, for example Fe, Cu and Co, which possess two or more valence states with a suitable oxidation–reduction potential affect both the speed of autoxidation and the direction of hydroperoxide breakdown to volatile compounds (Grosch, 1982). As has been discussed, transition metal ions in their lower valence state (M^{n+}) react very quickly with hydroperoxides. They act as one-electron donors to form an alkoxy radical and this can be considered as the branching of the propagation step. In a slow consecutive reaction, the reduced state of the metal ion may be regenerated by hydroperoxide molecules. Owing to the presence of water, or of metal complexing with chain termination products, the regeneration is not complete. The radicals produced enter the propagation sequence and decrease the induction period.

The catalytic activity of heavy metals depends in reality not only on the ion species and its redox potential but also on the ligands attached to it, the solvent system, the presence of electron donors such as ascorbate and cysteine which keep the metal ion in its lower valance state, and the pH. Maximum degradation of peroxides occurs in the pH region 5.0–5.5 and the activity for catalysing degradation decreases from $Fe^{2+} > Fe^{3+} > Cu^{2+}$, as detailed by O'Brien (1985). Metals can abstract a hydrogen atom from the fatty acids themselves, but the ubiquitous presence of traces of hydroperoxides in oils is likely to ensure that hydroperoxide decomposition is the normal initiation reaction.

6.7 Off-flavours from volatile lipid molecules

Teranishi (1989) has suggested that the detection odour threshold is defined as the minimum physical intensity detection where the subject is not required to identify the stimulus but to detect the existence of the stimulus. Alternatively, the threshold value can be defined as the minimum concentration in which a pure compound can be perceived by 50% of the taste panellists (Hamilton, 2001). Odour (or flavour) threshold values are given in Table 6.7. Some of the dienals have very low odour threshold

Table 6.7 Flavour thresholds (adapted from Forss, 1972)

Chain length	Acids (ppm in water)	Aldehydes (ppm in water)	Methyl ketones (ppm in water)	δ-lactone (ppm in water)	γ-lactone (ppm in water)
2	5.4				
3		0.17	100		
4	6.8	0.07	25		
5		0.07	8.4		
6	5.4	0.015	0.25		18.0
7		0.031	0.25		0.52
8	5.8	0.047	0.25	0.57	0.04
9		0.045	0.25		0.09
10	3.5	0.007	0.25	0.16	0.15

Oxidative rancidity as a source of off-flavours   153

values, for example the value for *trans*-2, *trans*-4-nonadienal is 0.0005 ppm in water.

Flavour notes are somewhat subjective. Thus in Table 6.8 two different words are used to describe the C_7 methyl ketone. According to one set of workers this taste is blue cheesy whilst for another set of workers it is suggested that anty or spicy are better descriptors. The flavour of blue mould cheese is due to the breakdown of the lipids. The milk fat is hydrolysed to give some short chain saturated fatty acids, some of which are β-oxidised, then decarboxylated. On reduction of these products, fatty acids, methyl ketones and methyl alcohols are produced. Litman and Numrych (1977) have shown that each of the lipid classes has characteristic flavour notes. It is also clear that as the chain length in these compounds increases, the flavour becomes fatty. Thus in Table 6.8, the C_2 aldehyde is fresh and pungent, the C_3 aldehyde is fresh and milky, the C_6 aldehyde is fresh and green, the C_8 aldehyde is fresh and citrus and the C_{11} aldehyde is fatty.

6.8 Case study: lipid autoxidation and meat flavour deterioration

The primary mechanism for the degradation of desirable flavour in stored meats is lipid autoxidation. Lipids in muscle foods, particularly their phospholipid components, undergo degradation to produce a large number of volatile compounds. Their degradation leads to the formation of an array of secondary products such as aldehydes, hydrocarbons, alcohols, ketones, acids, esters, furans, lactones and epoxy compounds as well as polymers. These latter classes of compounds are flavour-active, particularly aldehydes, and possess low threshold values in the parts per million or even parts per billion levels. They are responsible for the development of warmed-over flavour (WOF), as coined by Tims and Watts (1958), and meat flavour deterioration (MFD) (e.g. Drumm and Spanier, 1991).

The degree of unsaturation of the acyl constituents of meat lipids primarily dictates the rate at which MFD proceeds. Unsaturated lipids are more susceptible. Autoxidation of meat lipids gives rise to a number of hydroperoxides which, in conjunction with the many different pathways possible, decompose to a large number of volatile compounds. However, other factors might also affect the oxidation of meat lipids and formation of WOF as well as shelf-life of products (Spanier *et al.*, 1988; Gray *et al.*, 1996). In chicken meat, lack of α-tocopherol is the main reason for MFD and formation of undesirable WOF products. However, cooked turkey meat, despite its higher content of unsaturated lipids, may not readily develop WOF because it contains endogenous α-tocopherol. In addition, the presence of heme compounds and metal ions may also hasten the oxidation of meat lipids. Furthermore, the presence of salt and other

Table 6.8 Flavour notes (adapted from (a) Litman and Numrych, 1977 and (b) Forss, 1972)

Chain length	2	3	4	5	6	7	8	9	10	11
Aldehydes (a)		fresh, milky			fresh, green		fresh, citrus			fatty
2-Enals (a)			sweet, pungent	sweet, green				sweet, fatty, green		sweet, fatty
Methyl Ketones (a)		pungent, sweet	solvent, sweet			blue cheesy				fatty, sweet
Methyl Ketones (b)		acetone	ethereal, unpleasant	fruity, bananas		anty, spicy	floral, green	fruity, floral, fatty	citrus	rosy, orange
γ-Lactones (a)			oily	creamy, tobacco	creamy, coconut		coconut		peachy	
γ-Lactones (b)			sweet, slight caramel		sweet, tobacco	tobacco, sweet, herbaceous	sweet, nutty, oily	creamy, nutty, coconut		peachy, fruity

ingredients used in cooking, such as onion, may also affect progression of MFD. It has been reported that aldehydes, generated from oxidation of lipids, react with thiols, a class of compounds in onion, such as propenethiol, to produce 1,1-bis-(propylthio)-hexane and 1,2-bis-(propylthio)-hexane. These compounds will definitely modify flavour profile of muscle foods (Ho et al., 1994). Formation of volatile aldehydes and other lipid degradation products results in the masking of a desirable meaty aroma from products.

While adipose tissues generally contain over 98% triacylglycerols (TAGs), phospholipids constitute a major portion of intramuscular lipids of muscle foods. The unsaturated fatty acids present in TAGs of red meat and poultry contain mainly oleic and linoleic acids. However, phospholipids contain a relatively higher proportion of linolenic and arachidonic acids. In seafoods, long chain omega-3 fatty acids such as eicosapentaenoic and docosahexaenoic acids are prevalent. Existing differences in the fatty acid constituents of phospholipids and triacylglycerols of muscle foods are primarily responsible for species differentiation in cooked samples of meat, as discussed earlier. Furthermore, depending on the type and proportion of unsaturated fatty acids in muscle foods, lipid autoxidation and flavour deterioration may proceed at different rates. In this respect, seafoods deteriorate much faster than chicken which is oxidised faster than red meats.

Transition metals such as iron, copper and cobalt may catalyse the initiation and enhance the propagation steps involved in lipid autoxidation. For example, Fe^{2+} will reductively cleave hydroperoxides to highly reactive alkoxy radicals which in turn abstract a hydrogen atom from other lipid molecules to form new lipid radicals. This reaction is known as hydroperoxide-dependent lipid peroxidation (Svingen et al., 1979). Morrissey and Tichivangana (1985) and Tichivangana and Morrissey (1985) have reported that ferrous ion at 1 to 10 ppm levels acts as a strong pro-oxidant in cooked fish muscles. Similarly, copper(II) and cobalt(II) were effective pro-oxidants. These observations are in agreement with the findings of Igene et al. (1979) who reported that iron ions were the major catalysts responsible for enhancement of autoxidation in muscle foods.

Furthermore, Shahidi and Hong (1991) demonstrated that the pro-oxidant effect of metal ions was more pronounced at their lower oxidation state and found that in the presence of chelators such as the disodium salt of ethylenediaminentetraacetic acid (Na_2EDTA) and sodium tripolyphosphate (STPP), the pro-oxidant effect of metal ions was circumvented.

Hemoproteins in meats are generally known for their pro-oxidant activity (Robinson, 1924; Younathan and Watts, 1959; Pearson et al., 1977; Igene et al., 1979; Rhee, 1988; Shahidi et al., 1988; Johns et al., 1989; Shahidi and Hong, 1991; Wettasinghe and Shahidi, 1997); some have also been reported to possess antioxidant properties (Ben Aziz et al., 1971; Kanner et al., 1984; Shahidi et al., 1987; Shahidi, 1989; Shahidi and Hong, 1991; Wettasinghe and Shahidi, 1997). The pro-oxidant activity of heme compounds arises, at least in part, from their decomposition upon cooking the meat and from the liberation of free iron. Meanwhile, nitric oxide derivatives of heme pigments, namely nitrosyl myoglobin and nitrosyl ferrohemochrome (or cooked cured-meat pigment, CCMP), are reported to have an antioxidant effect (e.g. Wettasinghe and Shahidi, 1997).

6.9 Case study: lipid oxidation in fish

The oxidation of polyunsaturated fatty acids (PUFAs)-containing lipids causes the development of off-flavours and aromas, often referred to as 'rancidity' in fish. The compounds giving rise to rancid flavours and aromas are volatile secondary oxidation products derived from the breakdown of these lipid hydroperoxides. The main intrinsic factors that will determine the rate and extent of rancidity development in fish are:

- lipid level and fatty acid composition of the lipids
- levels of endogenous antioxidants and endogenous oxidative catalysts.

External influencing factors include:

- oxygen concentration
- surface area exposed to atmospheric oxygen
- storage temperature
- processing procedures that lead to tissue damage.

The levels of lipid in fish flesh vary depending on species, ranging from lean fish (<2% total lipid) such as cod, haddock and pollack, to high lipid species (8 to 20% total lipid) such as herring, mackerel and farmed salmon. A total lipid level of 5% has been suggested as a cut-off point between low and medium fat fish. In addition to species variability, lipid levels vary with sex, diet, seasonal fluctuation and tissue. For example the dark muscle of mackerel was found to contain 20% lipid in comparison with 4% in the white muscle, while Atlantic herring can have seasonal variation of 1 to 25% total fat (Body and Vlieg, 1989). It is well established that oily fish are particularly susceptible to lipid oxidation and rancidity development because of the high content of PUFAs in their lipid, particularly the nutritionally important *n-3* fatty acids eicosapentaenoic acid (20:5 *n*-3) (EPA) and docosahexaenoic acid (22:6 *n*-3) (DHA). Again, the fatty acid profile of fish varies quite considerably between and within species and is also influenced by the factors already mentioned.

The oxidation of PUFAs requires an active form of oxygen because the reaction of PUFAs with ground state oxygen is spin restricted. The spin restriction is overcome by an activation reaction (initiation) involving a catalyst to initiate free radical chain reactions (propagation). The lipid hydroperoxides formed are unstable and subsequently break down to form volatile compounds which are associated with rancidity development. The volatiles produced during storage have been thoroughly characterised for a number of oily fish species including anchovy, Atlantic salmon and mackerel and the mechanism of breakdown of lipid hydroperoxides to produce volatiles has been reviewed (Triqui and Reineccius, 1995; Refsgaard et al., 1998; Refsgaard et al., 1999; Hsieh and Kinsella, 1989).

In fish muscle there are a number of potential catalysts and mechanisms that may be involved in the activation reaction to generate active oxygen species for lipid oxidation. These include non-enzymatic mechanisms involving haem proteins such as myoglobin, haemoglobin and cytochrome P450, free iron and enzyme initiators such as lipoxygenase (LOX) and cyloxygenase (COX), all of which have been extensively reviewed (Harris and Tall, 1994; Kanner, 1992; Aust and Svingen, 1982). In reality, some of these mechanisms are of minor importance. For example, although enzymes such as LOX have been reported in post-mortem flesh, the overall impact on rancidity development is still questionable. Also there is no strong evidence for the role played by singlet oxygen. This is especially true when light is not a factor during storage.

6.10 Conclusions: preventing off-flavours

Off-flavours have always been a problem in fatty foods. One simple way in which lipid foods have an off-flavour is for the foodstuff to absorb a volatile compound from the environment in which the food is kept. This is probably the easiest form of rancid formation to prevent. If the food is stored in a good container and at a low temperature, then this form of rancid off-flavour should be of minimal importance. As far as the oxidation of lipid-containing foods is concerned, it is not possible to prevent the decomposition of the hydroperoxide once it has been formed. It is usual to try to minimise the formation of the hydroperoxide. Use of antioxidants has been and will continue to be the favoured method of improving storage times and reducing the production of off-flavours before the food is consumed. In this sense, the changeover to natural antioxidants, for example tocopherols, tocotrienols, flavonoids, green tea catechins and the carnosic acid derivatives will continue apace (Pokorny et al., 2001).

Some other new lipids may require less antioxidant. Medium chain triacylglycerols are made from $C_{8:0}$ and $C_{10:0}$ fatty acids found in high lauric oils like coconut and palm kernel oil (Heydinger, 1999). Medium chain, triacylglycerols are metabolised differently from longer chain triacylglycerols, i.e. they behave like carbohydrates in being hydrolysed more rapidly and completely than longer chain triacylglycols and are absorbed more quickly. They have been used in medical nutrition products for patients with poor fat absorption syndrome and for premature babies. They are odourless, tasteless and colourless and have great oxidative stability. Heydinger (1999) has demonstrated that they have an Active Oxygen method (AOM) value of 300 to 500 hours at a temperature of 100 °C. This compares with a value of 19 hours for soyabean oil. In this area of lipid nutrition the off-flavours are likely to develop very slowly. The advent of agricultural biotechnology has produced high oleic acid sunflower oil, high oleic acid safflower oil, low linolenic canola oil, high oleic canola oil and high lauric acid oil. All of these oils except the last one have lower levels of polyunsaturated fatty acids than the original oils. Loh (1999) has reported that potatoes fried in low linolenic acid canola oil have flavour intensity of the five characteristics, i.e. fishy, buttery, painty, cardboard and waxy ranging from bland to 2.05 (where 15 is objectionable). These values were very similar to those obtained when frying was done in two commercial shortenings, a partially hydrogenated soyabean oil and a partially hydrogenated soy bean oil and cottonseed oil blend. These are two of the ways in which technologists, industrialists and product developers are attempting to minimise the off-flavours and taints which are so widespread in the food industry.

6.11 Sources of further information and advice

AKOH C C and MIN D B (1998), *Food Lipids: chemistry nutrition and biotechnology*, New York, Marcel Dekker.
ALLEN J C and HAMILTON R J (1994), *Rancidity in Foods*, 3rd edn, Glasgow, Blackie Academic and Professional.
FRANKEL E N (1998), *Lipid Oxidation*, Dundee, The Oily Press.
GUNSTONE F D (2001), *Structured and Modified Lipids*, New York, Marcel Decker.
MIN D B and SMOUSE T H (1985), *Flavor Chemistry of Fats and Oils*, Champaign, American Oil Chemists' Society.
MIN D B and SMOUSE T H (1989), *Flavor Chemistry of Lipid Foods*, Champaign, American Oil Chemists' Society.
POKORNY J, YANISHLIEVA N and GORDON M (eds), *Antioxidants in Food: practical applications*, Woodhead Publishing, Cambridge.
ROSSELL B (2001), *Oils and Fats: Vol 2, Animal carcass fats*, Leatherhead, Leatherhead Food Research Association.
SUPRAN M K (1978), *Lipids as a Source of Flavor*, Washington DC, American Chemical Society.
WIDLAK N (1999), *Physical Properties of Fats, Oils and Emulsifiers*, Champaign, American Oil Chemists' Society.

6.12 References

ALLEN J C (1994), 'Rancidity in dairy products', in Allen J C and Hamilton R J (eds), *Rancidity in Foods*, Glasgow, Blackie Academic and Professional, pp 179–89.
AUST S D and SVINGEN B A (1982), 'The role of iron in enzymatic lipid peroxidation', in Pryor W A (ed), *Free Radical Biology*, Vol. 5, Academic Press, pp 1–28.
BEN-AZIZ A, GROSSMAN S, ASCARELLI I and BUDOWSKI P (1971), 'Linoleate oxidation induced by lipoxygenase and heme proteins: a direct spectro photometric assay', *Anal Biochem*, **34**, 88–100.
BODY D R and VLIEG P (1989), 'Distribution of the lipid classes and eicosapentaenoic and docosahexaenoic acids in different sites in blue mackerel fillets', *J Food Sci*, **54**, 569–72.

BUTTERY R G (1989), 'Importance of lipid derived volatiles to vegetable and fruit flavor', in Min D B and Smouse T H (eds), *Flavor Chemistry of Lipid Foods*, American Oil Chemists' Society, Champaign, USA, pp 156–66.
CRAWFORD R V (1985), 'Patterns of refined fat usage and practical constraints', in Padley, F B and Podmore, J (eds), *The role of Fats in Human Nutrition*, Chichester, Ellis Horwood, pp 182–207.
DRUMM T D and SPANIER A M (1991), 'Changes in the content of lipid autoxidation and sulfur-containing compounds in cooked beef during storage', *J Agric Food Chem*, **39**, 336–43.
DUDD S J (1999), PhD Thesis, Bristol University, 1999.
FAN L L, TANG J Y and WOHLMANN A (1983), 'Investigation of dec-l-yne formation is cottonseed oil fried foods', *J Am Oil Chem Soc*, **60**, 1115–21.
FORSS D A (1972), 'Odour and flavor compounds from lipids', in Holman, R T (ed), *Progress in the Chemistry of Fats and Other Lipids*, Oxford, Pergamon Press, pp 181–258.
FRANKEL E N (1998), *Lipid Oxidation*, Dundee, The Oily Press.
FUJIMOTO K (1989), 'Flavor chemistry of fish oils', in Min D B and Smouse T H (eds), *Flavor Chemistry of Lipid Foods*, American Oil Chemists' Society, Champaign, USA, pp 190–6.
GARDINER H W (1989), 'How the lipoxygenase pathway affects the organoleptic properties of fresh fruit and vegetables', in Min D B and Smouse T H (eds), *Flavor Chemistry of Lipid Foods*, Champaign, American Oil Chemists' Society.
GRAY J I, GOMAA E A and BUCKLEY D J (1996), 'Oxidative quality and shelf-life of meats', *Meat Sci*, **41**, 8111–23.
GROSCH W (1982), 'Lipid degradation products and flavour', in *Food Flavours Part A*, Morton I D and Macleod A J (eds), Chap. 5.
HAMILTON R J (2001), 'Rancidity of animal fats and a comparison with that of vegetable oils', in Rossell J B (ed), *Oils and Fats, Volume 2, Animal carcass fats*, Leatherhead, Leatherhead Food Research Association, pp 123–46.
HAMILTON R J, KALU C, MCNEILL G P, PADLEY F B and PIERCE J H (1998), 'The effects of tocopherols ascorbyl palmitate, and lecithin on autoxidation of fish oil', *J Am Oil Chem Soc*, **75**, 813–22.
HARRIS P and TALL J (1994), in *'Rancidity in Foods'*, Allen J C and Hamilton R J (eds), Blackie Academic and Professional, pp 256–70.
HEYDINGER J A (1999), 'Physical properties of medium-chain triglycerides and applications in food', in Widlak N (ed), *Physical Properties of Fats, Oils and Emulsifiers*, Champaign, American Oil Chemists' Society, pp 220–26.
HO C-T, OH Y-C and BAE-LEE M (1994), 'The flavour of pork', in Shahidi F (ed), *Flavor of Meat an Meat Products*, Blackie Academic and Professional, Glasgow, pp 38–51.
HOLE M (2001), Storage stability (a) Mechanisms of degradation, in *Encyclopaedia of Food Sciences and Nutrition*, London, Academic Press.
HSIEH R J and KINSELLA J E (1989), 'Oxidation of PUFAs, mechnisms, products and inhibition with emphasis on fish', *Adv Food Nutrition Res*, **33**, 233–41.
IGENE J O, KING J A, PEARSON A M and GRAY J I (1979), 'Influence of haem pigments, nitrite, non-haem iron on development of warmed-over flavor in cooked meat', *J Agric Food Chem*, **27**, 838–42.
JOHNS A M, BIRKINSHAWA L H and LEDWARD D A (1989), 'Catalysts of lipid oxidation in meat products', *Meat Sci*, **25**, 209–20.
KANNER J (1992), *'Mechanisms of Nonenzymatic Lipid Peroxidation in Muscle Foods in Lipid Oxidation in Foods'*, ACS Symposium Series 500, pp 55–73.
KANNER J, HAREL S, SHAGALOVICH J and BERMAN S (1984), 'Antioxidative effect of nitrite in cured meat products: nitric oxide–iron complexes of low molecular weight', *J Agric Food Chem*, **32**, 512–15.

KELLARD B, BUSFIELD D M and KINDERLERER J L (1985), 'Volatile off-flavour compounds in desiccated coconut', *J Sci Food Agric*, **36**, 415–20.

KUHN H, WIESNER R, LANKIN V Z, NEKRASOV A, ALDER L and SCHEWE T (1971), 'Analysis of the stereochemisty of lipoxygenage-derived hydroxy polyenoic fatty acids by means of high performance liquid chromatography', *Anal Biochim*, **160** (24).

LITMAN I and NUMRYCH S (1977), 'The role lipids play in the positive and negative flavours of food', in Supran M K (ed), *Lipids as a Source of Flavor*, Washington, American Chemical Society Symposium Series, No. 75, pp 1–18.

LOH W (1999), 'Biotechnology and vegetable oils: first generation products in the marketplace', in Widlak, N (ed), *Physical Properties of Fats, Oils and Emulsifiers*, Champaign, American Oil Chemists' Society, pp 247–55.

MACFARLANE N (2001), 'Methods of protection of products of increasing quality an value' in Gunstone F D (ed), *Structure and Modified Lipids*, New York, Marcel Dekker, 37–74.

MILO C and GROSCH W (1993), 'Changes in the odorants of boiled trout (*Salmo fario*) as affected by the storage of the raw material', *J Agric Food Chem*, **41**, 2076–81.

MIN D B and LEE E C (1988), *Proc of 5th International Flavour Conference*, Chalkidiki, Greece, Charalambous G (ed), Amsterdam, Elsevier.

MORRISSEY P A and TICHIVANGANA J Z (1985), 'The antioxidant activities of nitrite and nitrosylmyoglobin in cooked meats', *Meat Sci*, **14**, 175–90.

O'BRIEN P J (1985), 'Intracellular mechanisms for the decomposition of a lipid peroxide. I. Decomposition of a lipid peroxide by metal ions, haem compounds and nucleophiles', *Can J Biochem*, **47**, 485–92.

PEARSON A M, LOWE J D and SHORLAND F B (1977), 'Warmed-over flavor in meat, poultry and fish', *Adv Food Res*, **23**, 1–74.

POKORNY J, YANISHLIEVA N and GORDON M (eds) (2001), *Antioxidants in Food: practical applications*, Woodhead Publishing Ltd, Cambridge.

REFSGAARD H H F, BROCKHOFF P B and JENSEN B (1998), 'Sensory and chemical changes in farmed Atlantic salmon (*Salmo salar*) during frozen storage', *J Agric Food Chem*, **46**, 3473–9.

REFSGAARD H H F, HAAHR A M and JENSEN B (1999), 'Isolation and quantification of volatiles in fish by dynamic headspace sampling and mass spectometry', *J Agric Food Chem*, **47**, 1114–18.

RHEE K S (1988), 'Enzymatic and nonenzymatic catalysis of lipid oxidation in muscle food', *Food Technol*, **42** (6), 127–38.

ROBINSON M E (1924), 'Hemoglobin and methemoglobin as oxidative catalysts', *Biochem J*, **18**, 255–64.

SHAHIDI F (1989), 'Flavor of cooked meats', in Teranishi R, Buttery R G and Shahidi F (eds), *Flavor Chemistry: Trends and Developments*, ACS Symposium Series 388, American Chemical Society, Washington DC, pp 188–201.

SHAHIDI F and HONG C (1991), 'Role of metal ions and heme pigments in autoxidation of heat-processed meat products', *Food Chem*, **42**, 339–46.

SHAHIDI F, RUBIN L J and WOOD D F (1987), 'Control of lipid oxidation in cooked ground pork with antioxidants and dinitrosyl ferrohemochrome', *J Food Sci*, **52**, 564–7.

SHAHIDI F, RUBIN L J and WOOD D F (1988), 'Stabilization of meat lipids with nitrite-free curing mixtures', *Meat Sci*, **22**, 73–80.

SIMPSON G (1999), PhD thesis, Liverpool John Moores University.

SPANIER A M, EDWARDS J V and DUPUY H P (1988), 'The warmed-over flavor process in beef: a study of meat proteins and peptides', *Food Technol*, **42** (6), 110–18.

SVINGEN B AM BUEGE J A, O'NEIL F O and AUST S O (1979), 'The mechanism of NADPH-dependent lipid peroxidation: propagation of lipid peroxidation', *J Bio Chem*, **254**, 5892–9.

TERANISHI R (1989), 'Development of methodology for flavor chemistry past, present and future,' in Min D B and Smouse T H (eds), *Flavor Chemistry of Lipid Foods*, Champaign, American Oil Chemists' Society, 3–25.
TICHIVANGANA J Z and MORRISSEY P A (1985), 'Metmyoglobin and inorganic metals as prooxidants in raw and cooked muscle systems', *Meat Sci*, **15**, 107–16.
TIMS M J and WATTS B M (1958), 'Protection of cooked meats with phosphates', *Food Technol*, **12** (5), 240–3.
TRIQUI R and REINECCIUS G A (1995), 'Changes in flavour profiles with ripening anchovy', *J Agric Food Chem*, **43**, 1883–9.
VAN OS C P A, VENTE M and VLIEGENTHART J F G (1979a), 'An NMR shift method for the determination of enantiomeric composition of hydroperoxides formed by lipoxygenase', *Biochim Biophys Acta*, **574**, 103–8.
VAN OS C P A, RIJKE-SCHILDER G P M and VLIEGENTHART J F G (1979b), '9 L_R-Linoleyl hydroperoxide: a novel product from the oxygenation of linoleic acid by type-2 lipoxygenase from soybeans and peas', *Biochim Biophys Acta*, **575**, 479–84.
WETTASINGHE M and SHAHIDI F (1997), 'Antioxidant activity of preformed cooked cured-meat pigment in a β-carotene/linoleate model system', *Food Chem*, **58**, 203–7.
YOUNATHAN M T and WATTS B M (1959), 'Relationships of meat pigments to lipid oxidation', *Food Res*, **24**, 728–34.

7

The Maillard reaction as a source of off-flavours

A. Arnoldi, University of Milan, Italy

7.1 Introduction

We do not only eat food to assure a regular daily intake of all the nutrients we need. Pleasure is another important objective and the food industry knows how important it is to assure a consistent sensory quality in the items it produces. Sometimes sensory pleasure even overtakes the need for nutrients. A typical example is coffee, which makes a negligible contribution to daily caloric intake. It may be argued that we drink coffee because we are looking for the stimulus of caffeine, which is certainly true, but does not explain the consumption of decaffeinated coffee.

The sensory impact of foods depends mainly on 'flavour'. This is an English word not easily translatable in some other languages, such as Italian, Spanish or French, meaning an overall impression including taste and that part of food aroma which is perceived inside the mouth. Probably the easiest way to understand the meaning and significance of flavour is to make it disappear by chewing food while keeping the nostrils closed with two fingers. The flavour of the foods we consume raw, for example fruits or salads, depends on odorous materials, for example terpenes, deriving from specific biosynthetic pathways. Their perception may be enhanced by the addition of salt or sugar and a little lemon or vinegar. However, the great part of the foods we consume undergoes some sort of thermal processing during preparation that completely transforms their flavours. Coffee is a good example: green coffee is practically devoid of any flavour, while the aroma of a good Italian *espresso* coffee may be perceived at the distance of several metres. Two important chemical processes are responsible for flavour transformations during food processing: lipid autoxidation (which is the topic of Chapter 6) and the Maillard reaction.

7.2 Mechanism of the Maillard reaction

With the term Maillard reaction (MR), or non-enzymatic browning, since colour formation is a major feature of MR, we see a cascade of complex competitive reactions involving on one side amino acids, peptides and proteins and, on the other, reducing sugars (Arnoldi, 2001). The first observations on this process were done 90 years ago by Maillard (1912). Subsequently, Amadori reported the formation of a rearranged stable product deriving from sugars/amino acids interaction, which was named Amadori rearrangement product (ARP) (Amadori, 1931). The corresponding compound from fructose was described only 30 years after that by Heyns and Noack (1962). The first overall picture of the MR was proposed by Hodge (1953) and one of the most detailed descriptions of the pathways that lead to the main Maillard reaction products (MRPs) can be found in the excellent review by Ledl and Schleicher (1990).

Reducing sugars are indispensable in the MR. Pentoses, such as ribose, arabinose or xylose, although generally not very abundant in foods, are very effective in non-enzymatic browning, whereas hexoses, such as glucose or fructose, are less reactive, and reducing disaccharides, such as maltose or lactose, react rather slowly. Sucrose, as well as bound sugars (e.g. glycoproteins, glycolipids and flavanoids), are involved only after hydrolysis, induced by heating or, very often, by fermentation, as in dough leavening or cocoa bean preparation before roasting. Other essential compounds in MR are proteins or free amino acids already present in the raw material or produced in their turn by fermentation. In some cases (e.g. cheese) biogenic amines can react as amino compounds. Ammonia may be produced from amino acids during the MR to produce the particular kind of caramel colouring associated with MR.

Figure 7.1 presents a very simplified general picture of the MR. Following the classical interpretation by Hodge (1953), the initial step is the condensation of the carbonyl group of an aldose with an amino group to give an unstable glycosylamine (**1**), which undergoes a reversible rearrangement to the ARP (Amadori, 1931), i.e. a 1-amino-1-deoxy-2-ketose (**2**) (Fig. 7.2). The corresponding reaction of fructose gives, instead, a 2-amino-2-deoxy-2-aldose (**3**) (Heyns and Noack, 1962). A detailed description of the synthetic procedures, physicochemical characterisation, properties and reactivity of the ARPs may be found in an excellent review by Yaylayan and Huyggues-Despointes (1994).

In proteins the most relevant effect of the Maillard reaction is non-enzymatic glycosylation, which involves mostly lysine. The first glycosylation products are then converted to the Amadori product fructosyl-lsine

EARLY STAGE
First interactions between sugars and amino groups
INTERMEDIATE STAGE
Fissions, cyclisations, dehydrations, condensations, oligomerisations
Flavours and off-flavours
ADVANCED STAGE
Polymerisations ⇒ melanoidins

Fig. 7.1 Simplified scheme of the Maillard reaction.

R–NH2
protein or amino acid
1-amino-1-deoxyaldose (1)
1-amino-1-deoxyketose (2)
Amadori rearranged product

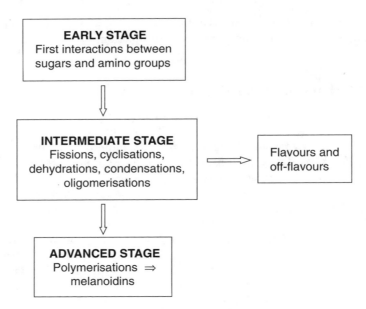

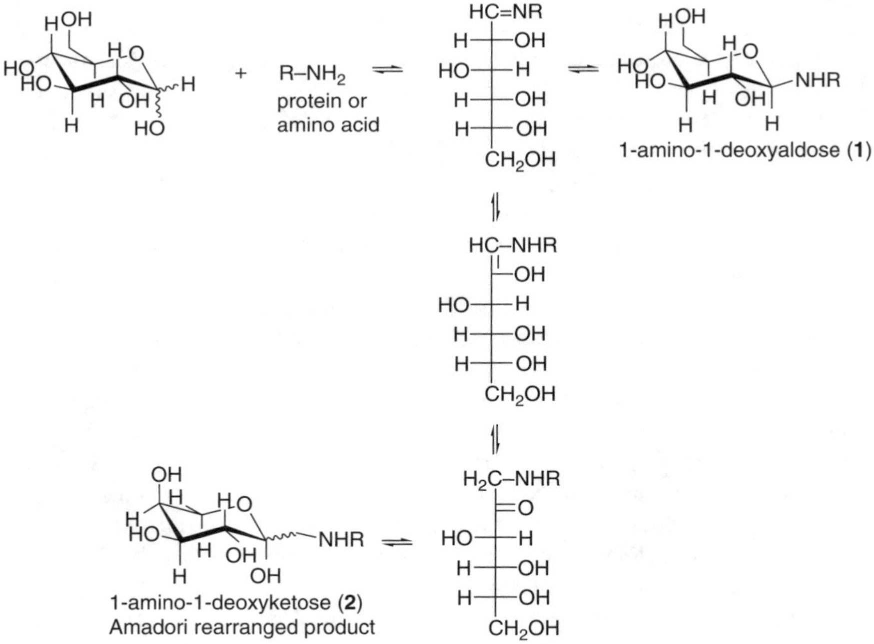

Fig. 7.2 Amadori rearrangement.

that can cross-link with adjacent proteins or with other amino groups. The resulting polymeric aggregates are called advanced glycation end products (AGEs). The formation of these compounds, which have been separated from model systems as well as from foods, takes place easily even at room temperature and is very well documented. It can be seen, for example, in some diseases, where long-lived body proteins and enzymes can be modified by glucose or other reducing sugars through the formation of ARPs (a process known as glycation) with subsequent impairment of many physiological functions, especially in diabetic patients and during ageing (Baynes, 2000; Furth, 1997; James and Crabbe, 1998; Singh *et al.*, 2001; Sullivan, 1996).

With low water contents and pH values in the range of 3 to 6, ARPs are considered to be the main precursors of reactive intermediates in model systems, whereas below pH 3 and above pH 8 or at temperatures above 130°C (caramelisation), sugars also degrade in the absence of amines (Ledl and Schleicher, 1990). Ring opening followed by 1,2 or 2,3-enolisation are crucial steps in the degradation of ARPs and are followed by dehydrations and fragmentations with the formation of many very reactive dicarbonyl fragments. This is considered to be the intermediate stage of the MR.

One of the first observations of Maillard was the production of CO_2, which is explained by the Strecker degradation (Fig. 7.3). An amino acid reacts with an α-dicarbonyl compound to produce an azovinylogous β-ketoacid (**4**) that undergoes decarboxylation. By this process amino acids are converted to aldehydes containing one less carbon atom, which are very reactive and often have very peculiar sensory properties, not always

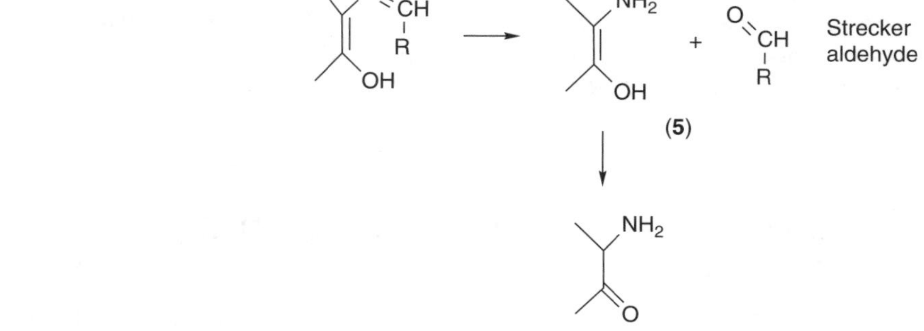

Fig. 7.3 Strecker degradation.

particularly pleasant. The most important consequence of the Strecker reaction is the incorporation of nitrogen in very reactive low-molecular-weight compounds deriving from sugars, such as (5), which are intermediates in the formation of very important flavours, such as pyrazines. However, the Strecker degradation has other important consequences, such as the formation of many odorous sulfur compounds, because the aldehydes deriving from cysteine and methionine degrade further to give hydrogen sulfide, 2-methylthiopropanal and methanethiol, which are eventually incorporated in some MRPs (Whitfield, 1992).

Since the early 1980s, other mechanisms not involving ARPs have been proposed. As an example, on the basis of the experimental observation of free radical formation in the very beginning of the MR, Hayashi and Namiki (1981, 1986) proposed a reducing sugar degradation pathway that produces glycolaldehyde alkylimines without passing through the formation of ARPs. In addition, taking into account experiments showing that sugars and most amino acids also undergo independent degradation (Yaylayan and Keyhani, 1996), Yaylayan (1997) has proposed a new conceptual approach to the MR. He has suggested that it is useful to define a sugar fragmentation pool {S}, an amino acid fragmentation pool {A} and an interaction fragmentation pool {D}, deriving from the Amadori and Heyns compounds (Table 7.1). Together they constitute a primary fragmentation pool of reactive building blocks that react to give a secondary pool of interaction intermediates and eventually a very complex final pool of stable end products.

Depending on food composition and process features, the MR produces thousands of different end products, which may be classified according to their role in foods. Thus very volatile compounds, such as pyrazines, pyridines, furans, thiophenes, thiazoles, thiazolines and dithiazines, are relevant for aroma, whilst some low molecular weight compounds are relevant for taste (Frank *et al.*, 2001; Ottiger *et al.*, 2001). Others behave as antioxidants, a few are mutagenic (Jägerstad *et al.*, 1998), whereas the brown polymers named melanoidins, the major MRPs in some foods such as coffee, roasted cocoa beans and malt or soy sauce, are responsible for food colour (Arnoldi *et al.*, 2002). The nutritional consequences of the MR have been reviewed (Arnoldi, 2002). This chapter will take into consideration only compounds relevant for flavour, which, although produced in very small yields during the MR, are known about in great detail since they are relatively easy to detect by GC-MS.

7.3 Relevant Maillard reaction products (MRPs) in food flavour

Nursten (1980–81) has suggested dividing flavour compounds into three groups that roughly resemble the Yaylayan classification (Yaylayan, 1997):

Table 7.1 Composition of the primary fragmentation pools

Amino acid fragmentation pool {A}
Amines
Carboxylic acids
Alkanes and aromatics
Aldehydes
Amino acid specific side chain fragments: H_2S (Cys), CH_3SH (Met), styrene (Phe)

Sugar fragmentation pool {S}
C1 fragments: formaldehyde, formic acid
C2 fragments: glyoxal, glycolaldehyde, acetic acid
C3 fragments: glyceraldehyde, methylglyoxal, hydroxyacetone, dihydroxyacetone, etc.
C4 fragments: tetroses, 2,3-butanedione, 1-hydroxy-2-butanone, 2-hydroxybutanal, etc.
C5 fragments: pentoses, pentuloses, deoxy derivatives, furanones, furans
C6 fragments: pyranones, furans, glucosones, deoxyglucosones

Amadori and Heyns fragmentation pool {D}
C3-ARP/HRP[a] derivatives: glyceraldehyde–ARP, amino acid–propanone, amino acid–propanal, etc.
C4-ARP/HRP derivatives: amino acid–tetradiuloses, amino acid–butanones
C5-ARP/HRP derivatives: amino acid–pentadiuloses
C6-ARP/HRP derivatives: amino acid–hexadiuloses, pyrylium betaines

Lipid fragmentation pool {L}
Propanal, pentanal, hexanal, octanal, nonanal
2-Oxoaldehydes (C6-9)
C2 fragments: glyoxal
C3 fragments: malondialdehyde, methylglyoxal
Formic acid, acids

[a] ARP = Amadori rearrangement product; HRP = Heyns rearrangement product.

1 Sugar dehydration/fragmentation products, such as furans, pyrones, cyclopentenes, carbonyls, dicarbonyls, acids;
2 Amino acids degradation products, Strecker aldehydes and sulfur compounds;
3 Volatiles produced by other interactions: pyrroles, pyridines, imidazoles, pyrazines, oxazoles, thiazoles and aldol condensation products.

A complete description of the pathways leading to these Maillard reaction products (MRPs) may be found in the review by Ledl and Schleicher (1990). Most of these compounds appear in the headspace of processed foodstuffs. However, their actual impact on flavour depends not only on their concentration, but also on their specific sensory thresholds, which are distributed across a wide range of values (see Table 7.2). The sensory characteristics of most MRPs may be found in a compilation by Fors (1983). As an example, Fig. 7.4 reports the threshold ranges of the main classes of volatile MRPs in meat flavour (Manley, 1994).

168 Taints and off-flavours in food

Table 7.2 Odour or flavour thresholds of some main Maillard reaction products (Fors, 1983)

Main ring	Substituents	Description		Threshold in water (ppb)
Furan	2-Methyl	Ethereal, sickly	O:	3 500
Furan	2-Pentyl	Fruity, sweet	O:	6 000
Furan	2-Hydroxymethyl	Slightly caramellic, sweet	F:	5 000
Furan	2-Carboxaldehyde	Pungent, sweet bread-like	O:	3 000
			F:	5 000
Furan	5-CH_2OH-2-CHO	Hay-like, warm-herbaceous	F:	100 000
Furan	2-Methanethiol	Roasted-coffee-like	F:	40
2(5H)-Furanone	3-Hydroxy-4-ethyl-5-methyl	Typically caramel	O:	500
3(2H)-Furanone	2,5-Dimethyl-4-methoxy	Reminiscent of sherry	O:	30
3(2H)-Furanone	4-Hydroxy-2,5-dimethyl (furaneol)	Fruity, caramel, burnt	O:	40
			F:	1 000
4H-Pyran-4-one	2-Methyl-3-hydroxy	Caramellic, sweet, fruity	F:	20 000
			O:	35 000
Pyrrole	2-CH_2COCH_3	Bready green	O:	10
Pyridine	2-Acetyl	Popcorn-like	O:	19
Pyrazine	2-Ethyl	Nutty, buttery	O:	22
Pyrazine	2-Isobutyl	Greeny, fruity	O:	400
Pyrazine	2,5-Dimethyl	Potato chips, nutty	O:	1
Pyrazine	2,6-Dimethyl	Chocolate	O:	9
Pyrazine	2-Ethyl-3-methyl	Nutty, roasted	O:	130
Pyrazine	2-Methyl-6-propyl	Burnt, butterscotch	O:	100
Pyrazine	2-Isobutyl-3-methyl	Roasted hazelnut	O:	35
Pyrazine	2,3-Dimethyl-5-ethyl	Chocolate, sweet	O:	1
Pyrazine	2,5-Dimethyl-3-ethyl	Roasted, nutty	O:	5
Pyrazine	2-Methoxy-3-methyl	Hazelnut, almond	O:	4
Pyrazine	2-Ethyl-3-methoxy	Raw potato	O:	0.4
Pyrazine	2-Isopropyl-3-methoxy	Bell pepper, earthy	O:	0.006
Pyrazine	2-Isopropyl-5-methoxy	Galbanum note	O:	10
Pyrazine	2-Isobutyl-3-methoxy	Bell pepper	O:	0.02
Pyrazine	2-Isobutyl-5-methoxy	Bell pepper	O:	0.016
Pyrazine	2-sec-Butyl-3-methoxy	Bell pepper	O:	0.1
Thiophene	2,5-Dimethyl	Fried onion	O:	1.3
Thiophene	2-Acetyl	Onion-like	F:	0.08
Thiophene	5-Methyl-2-CHO	Cherry-like	F:	1 000
1,2,4-Trithiolane	3,5-Dimethyl	Onion-like	F:	10
1,3,5-Trithiane	3-Methyl	Sulfurous	F:	0.04
Thiazole	2-Isobutyl	Spoiled wine-like	O:	2
Thiazole	2-Acetyl	Popcorn-like	O:	10
Thiazole	4-Ethyl-5-propyl	Roasted nutty	O:	0.06

O: odour threshold.
F: flavour threshold.

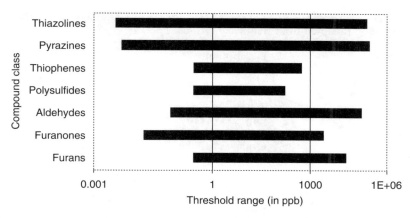

Fig. 7.4 Ranges of odour threshold values of main impact compounds in cooked meat.

A major problem facing technologists is how to determine which compounds are significant for flavour among thousands of MRPs. A possible solution is odour activity values (OAVs), a dimensionless number corresponding to the ratio of each volatile concentration in the food to its odour threshold. The underlying assumption is that the magnitude of OAV is a predictor of the flavour significance of a single compound in a very complex mixture. This assumption, of course, does not take into account any possible synergism. The evolution of this concept has produced two largely applied methodologies based on GC-sensory analysis: the CHARM analysis (Acree *et al.*, 1984) and the aroma extraction dilution analysis (Schieberle and Grosch, 1989). Schieberle *et al.* (2000) have tried to characterise potent aroma compounds in Maillard-type reactions using the OAV concept and to find a correlation with the concentration of precursors in foods.

Another very important reaction in foods involves lipids that can degrade by autoxidation giving in their turn reactive intermediates, mainly saturated or unsaturated aldehydes or ketones, but also glyoxal and methylglyoxal (in common with the Maillard reaction) and malondialdehyde (Table 7.1). This group can be classified as an additional lipid fragmentation pool {L} (D'Agostina *et al.*, 1998). The main aldehydes from oleic acid are octanal and nonanal, and from linoleic acid are hexanal, (E)-2-heptenal, (Z)- and (E)-2-octenal, (E,Z)- and (E,E)-2,4-decadienal, whereas linoleic acids gives a complex mixture very rich in (E,Z)-2,4-heptadienal (Belitz and Grosch, 1999). Their threshold values are reported in Table 7.3.

Clear interconnections between the MR and lipid autoxidation have been extensively studied in the case of food aromas, where many end products deriving from lipids and amino acids or sugars are very well

Table 7.3 Sensory properties of some selected aroma components from lipid peroxidation (adapted from Belitz and Grosch, 1999)

Compound	Description	Thresholds (ppm)	
		Nasal	Retronasal
Hexanal	Tallowy, green leafy	320	75
Octanal	Oily, soapy, fatty	320	50
Nonanal	Tallowy, soapy-fruity	13 500	260
Decanal	Orange peel-like	6 700	850
(Z)-Hexenal	Green-leafy	14	3
(E)-2-Heptenal	Cream, putty	2	1
(Z)-2-Octenal	Walnut	—	50
(E)-2-Octenal	Fatty, nutty	7 000	125
(Z)-2-Nanenal	Fatty, green-leafy	4.5	0.6
(E,Z)-2,4-Heptadienal	Frying odour, tallowy	4 000	50
(E,E)-2,4-Heptadienal	Fatty, oily	10 000	460
(E,Z)-2,6-Nonadienal	Cucumber-like	4	1.5
(E,Z)-2,4-Decadienal	Frying odour	10	—
(E,E)-2,4-Decadienal	Frying odour	180	40

documented (Whitfield, 1992). The mechanisms of formation of these compounds have very often been studied in model systems. Thus hexanal reacts with ammonium sulfide to give a variety of odorous compounds containing alkyl chains (Hwang *et al.*, 1986) that have been detected in various foods, particularly boiled allium, beef broth and cooked meat (Elmore and Mottram, 2000; Elmore *et al.*, 1997; Mottram, 1998). The identification of long-chain alkyl-substituted pyrazines in some cooked foods, particularly fried potatoes and baked and extruded corn-based products (Bruechert *et al.*, 1988), prompted an investigation into the reaction of fatty aldehydes with 1-hydroxypropanone and ammonium acetate (Chiu *et al*, 1990).

7.4 Food staling and off-flavours in particular foods

Food flavour is not stable and its deterioration is a major matter of concern to the food industry. Food staling can be described as a change in the aroma profile caused by loss of low boiling compounds and degradation processes. Storage temperature, penetration of oxygen in the package and loss of volatiles through diffusion are crucial factors when considering rancidity, and moisture can accelerate the staling process. Lipid autoxidation is generally considered to be the main source of off-flavours in food and many MRPs are considered to have a role in its prevention given their antioxidant properties (Alaiz *et al.*, 1997; Antony *et al.*, 2000, Lignert and Eriksson, 1981). However, in specific cases the MR is also a source of off-flavours. These negative aspects of the MR either during food processing or storage have been investigated much less frequently than lipid autoxidation.

From this point of view it is useful to divide food items into two groups:

1. Those in which quality means a flavour identical to unprocessed food, although thermal treatments are required for microbiological stabilisation.
2. Those in which the MR is intrinsically necessary for obtaining the typical flavour and texture.

Fruit juices are a typical example of the former class, in which a slow MR at room temperature during storage is often very detrimental to flavour quality. Browning and flavour deterioration of citrus fruit has been a problem throughout the history of the processing industry (Handwerk and Coleman, 1988). Maintaining the product at low temperature is still the major means of avoiding colour and flavour deterioration, with flavour usually becoming unacceptable before a detectable colour change. Although the low pH of these beverages is not particularly favourable for the MR, at least 14 out of 21 compounds detected in the flavour of old orange juice clearly derive from the Maillard reaction, for example 5-methylfurfural, furfural, 5-hydroxymethylfurfural, 2-(hydroxyacetyl)furan, 2-acetylfuran, 2-acetylpyrrole and 5-methylpyrrole-2-carboxaldehyde (Handwerk and Coleman, 1988). The presence of free amino acids in the juice has a relevant role in this phenomenon and their removal by ion-exchange resins has been proposed to increase stability. General information about the flavour changes during processing and storage of fruit juices may be found in a review by Askar (1999).

Milk processing is necessary to assure microbiological safety and an acceptable shelf-life, but it is also detrimental to flavour quality. The well known difference between the flavour and taste of pasteurised and UHT-treated milk is related to the Maillard reaction, and has encouraged research into milder technologies in milk processing. The Maillard reaction is also responsible for the deterioration of milk flavour during storage. Comparison of the flavour of different commercial UHT milk samples, either whole or skimmed, during four months, showed that the sensory characteristics of the latter were slightly worse and that many of the new components were related to both proteolysis and the MR (Valero et al., 2000). Information about the formation of off-flavours from the MR in whey protein concentrates may be found in a review by Morr and Ha (1991) and their formation in milk powder in a paper by Renner (1988). Imitation milk, such as soy milk, has also been investigated in detail (Kwok and Niranjan, 1995). Study of the deterioration of beer flavour by chromatographic olfactometry techniques has also shown that some compounds responsible for beer ageing are related to the MR (Evans et al., 1999).

Roasted products belong to a class of food items in which thermal processing is a requirement. An interesting paper by Warner et al. (1996)

has considered the problem of flavour fade and off-flavour formation in ground roasted peanuts, which is a major problem in the confectionary industry. Peanut oil is rich in polyunsaturated fatty acids, linoleic acid in particular, and the formation of fatty aldehydes, mainly hexanal, is significant in flavour formation. The authors monitored selected pyrazines, such as methylpyrazine, 2,6-dimethylpyrazine and 2,3,5-trimethylpyrazine, which are key impact compounds in roasted peanut flavour, and some aldehydes produced by lipid autoxidation, such as pentanal, hexanal, heptanal, octanal and nonanal. This investigation demonstrated that the concentration of pyrazines remains practically constant, whereas the concentration of aldehydes increases very quickly (e.g. the concentration of hexanal increases ten-fold in 68 days). Sensory evaluation confirmed the major contribution of lipid autoxidation to the formation of peanut off-flavour. This study is also significant for off-flavour formation during storage of other roasted nuts, such as filberts.

Another typical example of a thermally processed product is coffee. Cappuccio et al. (2001) have investigated the rate of staling in roasted and ground coffee at different temperatures after package opening and have tried to correlate the most significant chemical data and sensory evaluation. Among the most significant compounds are some MRPs, such as H_2S, methanethiol, 2-methylpropanal, diacetyl, 2-butanone, 2-methylfurane, 3-methyl- and 2-methylbutanal, 3-methylfuranthiol and 2-furfurylthiol. A decrease, especially in sulfur compounds, starts immediately after package opening, while compounds deriving from the degradation of lipids appear only after some days depending on the temperature.

Similar results have also been observed in the aroma of coffee brew, which changes shortly after preparation. Such observations are made in the manufacturing of instant coffee, heat sterilisation of coffee beverages or keeping freshly prepared coffee brews warm in a thermos flask. An investigation combining instrumental analysis with olfactometry perception has revealed a rapid decrease in the concentration of odorous thiols when coffee brews are stored or processed (Hofmann et al., 2001). The results have shown that, in particular, the key coffee odorant 2-furfurylthiol is significantly reduced (Hofmann and Schieberle, 2002), causing a strong decrease in the sulfury roasty odour quality in the overall aroma of the coffee beverages.

In an elegant experiment Hofmann et al. (2001) have shown that the addition of melanoidins isolated from coffee powder to an aqueous aroma recombinate, prepared using 25 coffee aroma compounds in the same concentration as determined in the original coffee brew, reduced the intensity of sulfury roasty odour quality in the headspace. Compounds particularly affected were 2-furfurylthiol, 3-mercapto-3-methylbutyl formate and 3-methyl-2-butene-1-thiol, known as the key thiols in coffee aroma. It was possible to demonstrate that these thiols were covalently bound to melanoidins via Maillard-derived pyrazinium compounds formed as oxida-

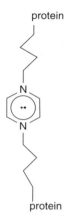

Fig. 7.5 Structure of CROSSPY.

tion products of 1,4-bis-(5-amino-5-carboxy-1-pentyl)pyrazinium radical cations (CROSSPY, Fig. 7.5). This is one of the first demonstrations that flavour staling may be due not only to the loss of key odorants and the formation of rancidity or undesired MRPs at room temperature, but also to the binding of specific compounds to food melanoidins or other polymers.

7.5 References

ACREE T E, BARNARD J and CUNNINGHAM D G (1984), 'A procedure for the sensory analysis of gas chromatographic effluents', *Food Chem*, **14**, 273–86.

ALAIZ M, HIDALGO F J and ZAMORA R (1997), 'Antioxidative activity of non-enzymatically browned proteins in oxidised lipid/protein reaction', *J Agric Food Chem*, **45**, 3250–4.

AMADORI M (1931), 'Condensation products of glucose with *p*-toluidine', *Atti R Accad Naz Lincei Mem Cl Sci Fis Mat Nat*, **13**, 72.

ANTONY S M, HAN I Y, RIECK J R and DAWSON P L (2000), 'Antioxidative effect of Maillard reaction products formed from honey at different reaction times', *J Agric Food Chem*, **48**, 3985–9.

ARNOLDI A (2001), 'Thermal processing and foods quality: analysis and control', in Richardson P (ed), *Thermal Technologies in Food Processing*, Cambridge UK, Woodhead Publishing Ltd, pp 138–59

ARNOLDI A (2002), 'Thermal processing and nutritional quality', in Richardson P. (ed), Woodhead Publishing Ltd, *The Nutrition Handbook for Food Processors*, Cambridge UK, 265–92.

ARNOLDI A, BOSCHIN G and D'AGOSTINA A (2002), 'Melanoidins in foods', *Res Adv Food Sci*, **3**, 1–10.

ASKAR, A. (1999), 'Flavor changes during processing and storage of fruit juices. Part 1. Markers for processed and stored fruit juices', *Fruit Process*, **9**, 236–44.

BAYNES J W (2000), 'From life to death – the struggle between chemistry and biology during aging: the Maillard reaction as an amplifier of genomic damage', *Biogerontology*, **1**, 235–46.

BELITZ H D and GROSCH W (1999), 'Lipids', in *Food Chemistry*, 2nd edn, Springer, Berlin, pp 152–236.
BRUECHERT L J, ZHANG Y, HUANG T C, HARTMAN Y G, ROSEN R T and HO C-T (1988), 'Contribution of lipids generated in extruded corn-based model systems', *J Food Sci*, **53**, 1444–7.
CAPPUCCIO R, FULL G, LONZARICH V and SAVONITTI O (2001), 'Staling of roasted and ground coffee at different temperatures: combining sensory and GC analysis', *Colloque Scientifique Int sur le Cafe* 19th, pp 151–61, and on CD.
CHIU E-M, KUO M-C, BRUECHERT L J and HO C-T (1990), 'Substitution of pyrazines by aldehydes in model systems', *J Agric Food Chem*, **38**, 58–1.
D'AGOSTINA A, NEGRONI M and ARNOLDI A (1998), 'Autoxidation in the formation of volatiles from glucose-lysine', *J Agric Food Chem*, **46**, 2554–9.
ELMORE J S, MOTTRAM D S, ENSER M and WOOD J D (1997), 'Novel thiazoles and 3-thiazolines in cooked beef aroma', *J Agric Food Chem*, **45**, 3603–7.
ELMORE J S and MOTTRAM D S (2000) 'Formation of 2-alkyl-(2*H*)-thiapyrans and 2-alkylthiophenes in cooked beef and lamb', *J Agric Food Chem*, **48**, 2420–4.
EVANS D J, SCHMEDDING D J M, BRUIJNJE A, HEIDEMAN T, KING B M and GROESBEEK N M (1999), 'Flavour impact of aged beers', *J Inst Brew*, **105**, 301–7.
FORS S (1983), 'Sensory properties of volatile Maillard reaction products and related compounds. A literature review', in Waller G R and Feather M S (eds), *The Maillard Reaction in Food and Nutrition*, ACS Symp. Ser. 215, American Chemical Society, Washington DC, pp 185–286.
FRANK O, OTTIGER H and HOFMANN T (2001), 'Characterization of an intense bitter-tasting 1*H*,4*H*-quinolizinium-7-olate by application of the taste dilution analysis, a novel bioassay for the screening and identification of taste-active compounds in foods', *J Agric Food Chem*, **49**, 231–8.
FURTH A J (1997), 'Glycated proteins in diabetes', *Br J Biomed Sci*, **54**, 192–200.
HANDWERK R L and COLEMAN R L (1988), 'Approaches to the citrus browning problem. A review.' *J Agric Food Chem*, **36**, 231–6.
HAYASHI T and NAMIKI M (1981), 'On the mechanism of free radical formation during browning reaction of sugars with amino compounds', *Agric Biol Chem*, **45**, 933–9.
HAYASHI T and NAMIKI M (1986), 'Role of sugar fragmentation in early stage browning of amino-carbonyl reaction of sugars with amino acids', *Agric Biol Chem*, **50**, 1965–70.
HEYNS K and NOACK H (1962), 'Die Umzetzung von D-fructose mit L-Lysine and L-Arginin und deren Beiziehung zu nichtenenzymatischen Bräunungsreaktionen', *Chem Ber*, 720–7.
HODGE J E (1953), 'Chemistry of browning reactions in model systems', *J Agric Food Chem*, **1**, 928–43.
HOFMANN T, CZERNY M, CALLIGARIS S and SCHIEBERLE P (2001), 'Model studies on the influence of coffee melanoidins on flavor volatiles of coffee beverages', *J Agric Food Chem*, **49**, 2382–6.
HOFMANN T and SCHIEBERLE P (2002), 'Chemical interactions between odor-active thiols and melanoidins involved in the aroma staling of coffee beverages', *J Agric Food Chem*, **50**, 319–26.
HWANG S-S, CARLIN J T, BAO Y, HARTMAN G J and HO C-T (1986), 'Characterisation of volatile compounds generated from the reactions of aldehydes with ammonium sulfide', *J Agric Food Chem*, **34**, 538–42.
JÄGERSTAD M, SKOG K, ARVIDSSON P and SOLYAKOV A (1998), 'Chemistry, formation, and occurrence of genotoxic heterocyclic amines identified in model systems and cooked foods', *Z Lebensm-Unters-Forsch A*, **207**, 419–27.
JAMES M and CRABBE C (1998), 'Cataract as a conformational disease – The Maillard reaction, α-crystalline and chemotherapy', *Cell Mol Biol*, **44**, 1047–50.

KWOK K-C and NIRANJAN K (1995), 'Effect of thermal processing on soymilk', *Int J Food Sci Technol*, **30**, 263–95.
LEDL F and SCHLEICHER E (1990), 'New aspects of the Maillard reaction in foods and in the human body', *Angew Chem, Int Ed Engl*, **29**, 565–706.
LIGNERT H and ERIKSSON C E (1981), 'Antioxidative effect of Maillard reaction products', *Prog Food Nutrition Sci*, **5**, 453–66.
MAILLARD A C (1912), 'Action des acides amines sur les sucres. Formation des melanoidines par voie methodologique', *C R Acad Sci*, **154**, 66–8.
MANLEY C H (1994), 'Process flavour and precursors systems', in Parliment T H, Morello M J and McGorrin R J (eds), *Thermally Generated Flavors. Maillard, Microwave and Extrusion Processes*, ACS Symposium Series 543, American Chemical Society, Washington DC, pp 16–25.
MORR C V and HA E Y W (1991), 'Off-flavors of whey protein concentrates: a literature review', *Int Dairy J*, **1**, 1–11.
MOTTRAM D S (1998), 'Flavor formation in meat and meat products: a review', *Food Chem*, **62**, 415–24.
NURSTEN H E (1980–81), 'Recent developments in studies of the Maillard reaction', *Food Chem*, **6**, 263–77.
OTTIGER H, BARETH A and HOFMANN T (2001), 'Characterization of natural "cooling" compounds formed from glucose and L-proline in dark malt by application of taste dilution analysis', *J Agric Food Chem*, **49**, 1336–44.
RENNER E (1988), 'Storage stability and some nutritional aspects of milk powders and ultra-high temperature products at high ambient temperatures', *J Dairy Res*, **55**, 125–42.
SCHIEBERLE P and GROSCH W (1989), 'Bread flavor', in Parliment R J, McGorrin R J and Ho C T (eds), *Thermal Generation of Aromas*, ACS Symposium Series 409, American Chemical Society, Washington DC, pp 258–67.
SCHIEBERLE P, HOFMANN T and MUNCH P (2000), 'Studies on potent aroma compounds generated in Maillard-type reactions using the odor-activity-value concept', *Flavor Chemistry*, ACS Symposium Series 756, American Chemical Society, Washington DC, pp 133–50.
SINGH R, BARDEN A, MORI T and BEILIN L (2001), 'Advanced glycation end-products: a review', *Diabetologia*, **44**, 129–46.
SULLIVAN R (1996), 'Contributions to senescence-non-enzymatic-glycosylation of proteins', *Arch Physiol Biochem*, **104**, 797–806.
VALERO E, VILLAMIEL M, MIRALLES B, SANZ J and MARTINEZ-CASTRO I (2000), 'Changes in flavor and volatile components during storage of whole and skimmed UHT milk', *Food Chem*, **72**, 51–8.
WARNER K J H, DIMICK P S, ZIEGLER G R, MUMMA R O and HOLLENDER R (1996), '"Flavor-fade" and off-flavors in ground roasted peanuts as related to selected pyrazines and aldehydes', *J Food Sci*, **61**, 469–72.
WHITFIELD F B (1992), 'Volatiles from the interactions of the Maillard reaction and lipids', *Crit Rev Food Sci Nutrition*, **31**, 1–58.
YAYLAYAN V A (1997), 'Classification of the Maillard reaction: a conceptual approach', *Trends Food Sci Technol*, **8**, 13–18.
YAYLAYAN V A and HUYGGUES-DESPOINTES A (1994), 'Chemistry of Amadori rearrangement products', *Crit Rev Food Sci Nutrition*, **34**, 321–69.
YAYLAYAN V A and KEYHANI A (1996), 'Py/GC/MS analysis of non-volatile flavor precursors: Amadori compounds', in Pickenhagen W, Ho C-T and Spanier A M (eds), *Contribution of Low and Non-volatile Materials to the Flavor of Food*, Carol Stream IL, USA, Allured Publishing, pp 13–26.

8

Off-flavours due to interactions between food components

E. Spinnler, INRA, France

8.1 Introduction

In a food matrix different elements can generate off-flavours individually. However, some off-flavours can arise from the interaction of compounds in food during formulation or packaging. These compounds may derive from raw materials, additives, flavours or the packaging chosen. They react spontaneously or are catalysed by enzymes or micro-organisms. Reactions can also be stimulated by the manufacturing processes used.

This chapter will describe some of the different interactions that may be responsible for off-flavours. It begins with the interactions between the food matrix and flavour compounds. The chapter will then consider biological interactions with the food matrix and how these can produce off-flavours.

8.2 Flavour compound volatility in different food matrices

There is a large variety of interactions between flavour compounds and the food matrix. These interactions depend on the relationships between a number of components including:

- the solvent (water, fat)
- solute concentrations (sugar, salt, etc.)
- the macromolecular content of the food matrix (lipids, proteins or polysaccharides for example)
- the chemical properties of the flavour compounds.

The diversity of chemical structures in flavour compounds will determine the degree of retention by the food matrix. A change in the composition of flavour compounds will lead to a change in the interactions between them and the food matrix, producing changes in the flavour perceived. Changes produced by storing the food will also influence these interactions. In some food matrices and storage conditions the changes are negligible or even beneficial. In others the matrix may favour the release of undesirable flavour compounds due to lipid oxidation or other undesired reactions during food ageing.

In many cases, off-flavours are due to the abnormal concentration in the vapour phase of a normal flavour compound. Analysing the liquid/vapour equilibrium provides a means of quantifying the volatility of flavour compounds in different food matrices. The laws of thermodynamics mean that the equilibrium between vapour and liquid phases is determined by chemical potential. In a closed flask containing a diluted solution at a defined pressure and temperature, at equilibrium between both phases, this potential is minimum. Below its boiling temperature, if a volatile compound has strong affinity for a solvent, the chemical potential of the molecule will be lower in the solvent than in the vapour phase. As an example, a very hydrophobic flavour compound such as gamma-decalactone will be more soluble in oil than in water because the chemical potential in oil is much lower than in water. In the same way, at the same low concentration (e.g. 1 mg/l), the partial pressure of the volatile in an oil solution will be lower than in a water solution.

Since they are caused by interactions at the molecular level, the molar fraction is commonly used to describe these phenomena. The molar fraction is defined as the number of moles of the compound in the total number of moles (solutes and solvent) in a defined volume. In the vapour phase, a volatile compound is characterised by a molar fraction (y_i) and by a partial pressure p_i (Fig. 8.1). In a real solution:

$$p_i = \gamma_i \cdot x_i \cdot p^0_i(T) \qquad [8.1]$$

where P^0_i is the saturated vapour pressure of the compound i, at the temperature T, x_i is the molar fraction of i in the solvent and γ_i is the coefficient of activity of the compound i.

In very dilute solution:

$$\gamma_i = \gamma_i^\infty \qquad [8.2]$$

Experimentally γ_i^∞ is easily calculated as the reverse of the molar fraction at the limit of solubility:

$$\gamma_i^\infty = 1/x_i^{sat} \qquad [8.3]$$

and

$$p_i = k_i x_i \qquad [8.4]$$

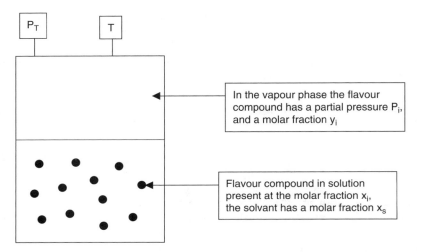

Fig. 8.1 Scheme of a simple equilibrium between a liquid phase and the vapour phase.

where k_i is the Henry's constant

with $\quad p_i = y_i . P_T \quad\quad\quad\quad\quad\quad\quad\quad\quad\quad\quad\quad\quad\quad\quad$ [8.5]

A key parameter in the vapour/liquid equilibrium is the partition coefficient liquid/vapour K_i. This coefficient allows us to quantify the affinity of the volatile compound for the food matrix. K_i is defined as the ratio between the molar fraction in the vapour to the molar fraction of i in the liquid:

$$K_i = y_i / x_i = \gamma_i P^0_i (T) / P_T \quad\quad\quad [8.6]$$

Another important parameter, when several flavour compounds are in a mixture, as is generally the case, is the relative volatility of a compound (i) as compared to another (j). This relative volatility $\alpha_{i/j}$ is the ratio of the partition coefficient K_i to the partition coefficient K_j:

$$\alpha_{i/j} = K_i / K_j = \gamma_i P_i^{sat} / P_j^{sat} \quad\quad\quad [8.7]$$

The relative volatility of compound i to water volatility may be expressed as follows:

$$\alpha_{i/w} = K_i / K_j = \gamma_i P_i^{sat} / P_w^{sat} \quad\quad\quad [8.8]$$

If the food matrix changes, this ratio will also change as compounds behave differently in different matrices. This will lead to change in the flavour

Table 8.1 Example of the behaviour of 5 compounds in water. P_i^0 is the saturated vapour pressure, Log(P) is the Logarithm of the partition ratio of the compound i between n-octanol and water, γ_i is the activity coefficient of the compound i at infinite dilution, k_i is a ratio of concentrations $k_i = C^{gas}/C^{water}$, K_i is the ratio of molar fraction between the gas and the liquid, $\alpha_{i/w}$ is the relative volatility of the compound i to water

Compound (molar mass)	Boiling temperature (°C)	P_i^0 (Pa)[a]	Solubility in water[a] (g/l)	Log(P)	γ_i^∞	k_i[a] (* 10⁵)	K_i	$\alpha_{i/w}$
Water (18)	100	4123	—		1			
Butyric acid (88)	163.7	217	60	0.79	82	2.18	0.18	0.0095
3-Octanone (128)	167.5	263[b]	2.6[b]	2.22	2735	531[b]	7	0.45
γ-decalactone (170)	280	0.6	0.65	3.35	14230	8.8	0.08	2
Diacetyl (86)	88	7474	20[c]	-1.34	239	54.3	18	33
Styrene (104)	145	842	0.310	2.95	18638	11240	157	32

[a] at 25°C.
[b] at 20°C.
[c] at 15°C.

perception of the product. As an example, Table 8.1 shows that the flavour compound γ-decalactone, which has a boiling point 2.8 times higher than water, is twice as volatile as water in a water-based solution because of its hydrophobicity which leads to a very high value activity coefficient. The structure of the flavour compound will also influence its behaviour. As an example, K_i is higher for alcohol, ketone, ester and aldehyde, depending on the carbon chain. In a homologous series, the longer the carbon chain, the higher is the hydrophobic character of the molecule.

8.3 Flavour retention in different food matrices

In a food matrix lipids have a particularly high retention capacity for flavour compounds. This retention capacity is very important for fat free products. As an example some fat free 'fromage frais' develop unexpected and undesirable flavours once the fats are removed. There are two main reasons for this change. First, the treatment of the milk to separate the fat may cause oxidation of the trace amount of fat still in the product after treatment. Second, as the product no longer contains much fat, all oxidation volatiles such as (Z) 2-nonenal, (Z) 4-heptenal, 1-Octen-3-one, which are quite hydrophobic and have high γ_i value, will be released very easily by the high moisture content of the matrix. This process generates undesirable cardboard or metallic flavours. Retaining a very small amount of fat (for example 1%) may help to avoid this problem without significantly changing the nutritional benefit of the low-fat food product. However, fats can increase the risk of taint by trapping chemicals from the food processing environment. Fats have been found to trap chlorophenols, chloroanisoles from packaging materials, or from wood or paint.

Other compounds have an important retention capacity. Proteins such as Bovine Serum Albumine, β-lactoglobuline, caseins and soya proteins have been shown to have a retention potential (Fares *et al.*, 1998). Polysaccharides such as starch have also a capacity for flavour retention. A recent study by Lopez da Silva *et al.* (2002) has shown that the retention of the flavour compounds nonanal, decanal and, to a smaller extent, (E) 2-nonenal and (E,E) 2,4 decadienal is significantly higher in a suspension (2% water) of nongelatinised starch than in water alone. A smaller effect is observed when starch is gelatinised. However, inulin, which is sometimes used as fat replacer, has a much smaller retention effect (Gijs *et al.*, 2000). Emulsifiers also affect flavour release. They do so by changing the rheology of the food matrix and the exchange surface of the interface between lipids and water. It has been shown that 2-stearoyl lactylate, for example, facilitates flavour release in a starch suspension (Lopez da Silva *et al.*, 2002). In contrast, small molecules such as salt or sugars may provoke a salting out effect and emphasise taints.

8.4 Off-flavours caused by reactions between components in the food matrix

In any food product, each ingredient should not cause off-flavours itself but, in the process of addition to the matrix, should not liberate other compounds giving unpleasant flavours. This rule also applies to environmental chemicals which might come into contact with a food product. These may cause taints themselves, but they can react with components in the food matrix to generate off-flavours (Reineccius, 1991).

A classical example is phenolic derivatives. They can originate from raw materials such as milk, meat, malt, fruits or vegetables. They can also originate from wood or cardboard in contact with the food before or after the packaging (O Connell and Fox, 2001). As an example, chlorophenols are sometimes found in milk products. They are produced in three main ways (de Jong et al., 1994):

- via wood treatment products, disinfectants or pesticides contaminating the food directly
- through spontaneous reaction between phenolic compounds present in food products and chlorine or other halogenated compounds largely used for sanitation
- by biological chlorination of phenols from chloride ions through haloperoxidase activity exhibited by different fungi.

Particular care should be taken in the use of chlorine. Chlorine combines easily with aromatic cycles of phenols, which are very common in foodstuffs, to produce chlorophenols. These compounds have a proven role in different taints and off-flavours. These abnormal phenol compounds have quite high olfactory thresholds. However, they are also very easily methylated by fungi to produce polychloroanisoles with much lower thresholds. They then become responsible for musty off-flavours (Bosset et al., 1994).

Other ingredients deliberately added to food may cause reactions which then produce off-flavours. Preservative agents can be a source of off-flavours. As an example, it has been shown that sulfites in wine are a stimulating factor for the synthesis of 2-aminoacetophenone responsible for 'untypical aging off-flavour' (UTA) (Hoenicke et al., 2002). The same mechanism could be responsible for stale flavour in beer (Palamand and Grigsby, 1974). Some antioxidants such as gallic acid can promote the decomposition of lipid hydroperoxides to volatile oxidation products. It has been shown that the small size of oil droplets increases the propensity to fat oxidation. Jacobsen et al. (2001) have shown that EDTA, for example, is a very efficient antioxidant but that it reduces the size of fat globules in mayonnaise. The appearance of oxidative off-flavours of this kind is an important problem for products such as mayonnaise, especially when enriched with omega-3 and omega-6 polyunsaturated fatty acids. These reactions can also be stimulated by particular food processes. Irradiation, for example, stimulates oxidation reactions. The use of antioxidants such as sesamol has been shown to be efficient in reducing off-flavours in pork patties (Chen *et al.*, 1999). However, neither vitamin E, sesamol, rosemary and gallic acid reduced effectively the off-flavours of irradiated turkey sausages (Du and Ahn, 2002).

For ionisable compounds pH is important because only the molecular form of the compound is volatile. As an example, butyric acid, which is an important compound involved in the flavour intensity of butter may, in certain circumstances, be responsible for a rancid off-flavour. It has been shown that the pH of the butter is a key influence on the behaviour of the butyric acid in milk-fat derived products. The pH of the butyric acid is 4.8. Below this pH most of the compound is in its molecular form and as a consequence more volatile. In traditional butters the pH can be around 4.6, and the higher volatility of the butyric acid will lead to butters with higher flavour intensities. Excessive release of butyric acid prompted by some production techniques causes a rancid off-flavour.

8.5 Bacterial interactions with the food matrix causing off-flavours

Some raw materials can be precursors of off-flavours in final food products because of certain micro-organisms. The case of fruit juices contaminated by recently identified acidophilic bacteria illustrates this. *Alicyclobacillus acidiphilus* is a novel thermo-acidophilic omega-alicyclic fatty acid-containing bacterium isolated from acidic beverages such as apple juice and orange juice. It is able to synthesise gaiacol which is widely responsible for off-flavours (Matsubara *et al.*, 2002). In brewing free fatty acids are associated with the formation of aldehydes which produce stale off-flavours. The level of free fatty acids is related to lipase activity (Schwartz *et al.*, 2002). Lipase and protease activities are also stimulating factors for cheese ripening. Protease can be a cause of bitterness (Alkhalaf *et al.*, 1988) and lipase the cause of soapy off-flavours (Kheadr *et al.*, 2002).

Bitterness is a major off-flavour in a large diversity of foodstuffs. The compounds responsible are very often polyphenols or peptides, for example in dairy products such as cheese (Lemieux and Simard, 1991). In camembert cheese, for example, the association of different strains of *Penicillium camemberti* and *Geotrichum candidum* can significantly change degrees of bitterness. (Molimard *et al.*, 1994). *Penicillium camemberti* significantly increases proteolytic activity which produces bitter peptides. In contrast, the use of *Geotrichum candidum*, which produce high levels of peptidases (carboxy-peptidases and amino-peptidases), results in cheeses perceived as less bitter.

The availability of substrates can be critical in how quickly bacteria catalyse off-flavours. As an example, mould ripened cheeses made using stabilised curd technology have sometimes developed a celluloid off-flavour (Adda *et al.*, 1989). *Penicillium camemberti* has been shown to be responsible for the production of the styrene causing this off-flavour (Spinnler *et al.*, 1992). When substrates like lactose or lactate are exhausted the *Penicillium* attacks proteins and fat. Amino acids such as the phenylalanine are then degraded to form styrene with cinnamic acid as the metabolic intermediate. Substrate uptake is more intense at the cheese surface, where *Penicillium* grows, than in the inner cheese. If the uptake of *Penicillium* is quicker than the diffusion of lactate from the inner cheese to the rind, as is the case for ripening temperatures over 15°C, the starving *Penicillium* starts to break down the other molecules of the medium such as fat or proteins. The reaction is also accelerated when curd is washed to remove a part of the lactose and lactate in order to speed up the ripening reactions.

The role of bacterial and enzymatic action in generating off-flavours can be complex. Tryptophan degradation has been identified as a factor in meaty and faecal off-flavours in cheddar cheese (Ummadi and Weimer, 2001) The process involves the ripening bacteria *Brevibacterium linens*. However, more than 14 enzyme activities are involved in tryptophan degradation by *Brevibacterium linens*, suggesting that *Brevibacterium linens* alone is not responsible.

In wine fermentation, wild yeasts such as *Saccharomyces cerevisiae*, *Rhodotorula sp.*, *Candida sp.*, *Cryptococcus sp*, *Pichia*, *Hansenula* and especially *Brettanomyces* are responsible for phenolic off-flavours (POF). These yeasts are able to decarboxylate ferulic acid into Vinyl 4 gaiacol or cinnamic acid into vinyl 4 phenol which cause the off-flavours. This defect is also observed in beer brewing. Yeasts are now selected for their low ability to produce POF, but it is also important to control phenolic acid content during wine production to reduce the risk of yeasts causing this reaction (Shinohara *et al.*, 2000).

8.6 Bacterial interactions with additives causing off-flavours

Additives may, in some cases, be responsible for off-flavours when in contact with enzymes and micro-organisms. As an example, a fishy off-flavour has been found in coffee cream (Eyer *et al.*, 1990). The intensity of the off-flavour was strongly pH dependent. A dynamic headspace extraction together with analysis by GC-MS identified trimethylamine and ethanol as the causes of the off-flavour. These compounds were caused by bacterial degradation of choline in lecithin during milk treatment. Lecithin from soy or rice is used as an additive in fermented milk products and has also been associated with oxidation off-flavours. The oxidation was attributed to the production of hydrogen peroxide by *Lactococcus lactis* ssp *lactis* (Suriyaphan *et al.*, 2001). Hydrogenated soy lecithin or lecithin with a small content of polyunsaturated fatty acids are not sensitive to this oxidation process.

Sorbic acid, which is used to prevent growth of moulds and yeasts, is also involved in off-flavours. Horwood *et al.* (1981) attributed a kerosene off-flavour found in Feta cheese to the flavour compound 1,3-pentadiene. This compound was produced from the decarboxylation of sorbic acid ((E,E)-2,4-hexadienoic acid) by bacterial activity.

Some flavour additives may interact with enzymes and components in the food matrix to produce off-flavours. It has been clearly established for example that milk proteins are able to interact with vanillin or other phenolic compounds to produce off-flavours. Two explanations have been advanced:

- an interaction of vanillin to cystein residues through a cysteine–aldehyde condensation reaction and/or Schiff base formation
- hydrophobic interactions (Reiners *et al.*, 2000).

Recently, it was shown that the flavour of vanilla-flavoured ice cream was quite unstable. Storage of the ice cream led to the appearance of a cardboard off-flavour. This off-flavour is often attributed to fat autoxidation or packing material. However, an enzyme present in the milk, xanthine oxidase, produces hydrogen peroxide and superoxide radicals from the conversion of vanillin into vanillic acid in the presence of oxygen. These two compounds produce perhydroxyl radicals which react with unsaturated fatty acids to form peroxides which produce the flavour compounds (E)-2-nonenal or heptanal responsible for a cardboard off-flavour (Gassenmeier, 2002, see Fig. 8.2). A sufficient heat treatment of the ice cream premix in denaturating the xanthine oxidase is an efficient way of preventing the appearance of the cardboard off-flavour.

8.7 Conclusion: identifying and preventing off-flavours

The increasing need of quality control in the food industry is a real challenge when the chemical and structure complexity of the foodstuffs is considered. This chapter illustrates how combinations of raw materials, additives, bacterial activity or production processes can interact to produce off-flavours. Unexpected reactions can occur during the development of new recipes, for example, producing off-flavours which are difficult to explain.

When an off-flavour occurs in a foodstuff, the starting point in an investigation is the identification of the compounds causing the off-flavour. Three types of analysis can be considered:

- sensory analysis
- chemical analysis
- microbiological analysis.

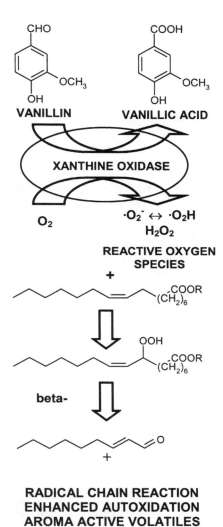

Fig. 8.2 Proposed pathway for the formation of (E)-2-nonenal and fatty acid hydroperoxides in vanilla ice cream.

Sensory analysis (discussed in Chapter 2) gives quick results and, if the off-flavour is familiar, an expert can suggest the possible compounds that might be responsible. This hypothesis will help direct further instrumental analysis. Instrumental methods are discussed in detail in Chapter 3 of this book. It is important, in this respect, to conduct both a chemical and microbiological analysis of the foodstuff. As this chapter has shown, bacteria can play an important role in catalysing chemical reactions which produce off-flavours.

Once this is done, it is necessary to identify how those compounds arose. Two immediate avenues of inquiry are:

- identifying the changes in product formulation or processing operations
- a comparison of the defective product with examples of products with a normal flavour.

These approaches will help narrow the field of enquiry. However, to take the investigation further, the analyst will then need a good understanding of the structure and functionality of the ingredients suspected of being responsible for the off-flavour and the ways they can interact within the food matrix. This chapter has tried to describe and categorise some of these interactions. It has suggested a means of identifying the volatility of flavour compounds and the retention capacity for such compounds of key ingredients. It has also shown how ingredients such as preservatives can interact with other components in the food matrix to generate off-flavours, and how these interactions can be catalysed or accelerated by factors such as pH or bacterial action. A deeper understanding of these interactions will help analysts and product developers to identify and avoid these problems more effectively in the future.

8.8 References

ADDA J and DEKIMPE J (1989), Production de styrène par *Penicillium camemberti* Thom. *Lait* **69**, 115–20.

ALKHALAF W, PIARD J C, EL SODA M, GRIPON J C, DESMAZEAUD M J and VASSAL L (1988), Liposomes as proteinase carriers for the accelerated ripening of Saint Paulin type cheese. *J. Food Sci.* **53**, 1674–79.

BOSSET J O, SIEBER R and SCHMUTZ F (1994), Occurrence and origin of chlorophenols in cheese rind: a review. *Schweizerische Milchwirtschaftliche Forschung*, **23**, 47–52.

CHEN X, JO C, LEE J L and AHN D U (1999), Lipid oxidation, volatiles and color changes of irradiated pork patties as affected by antioxidants. *J. Food Sci.* **64**, 16–19.

DE JONG E, FIELD J A, SPINNLER H E, WIJNBERG J B P A and DE BONT J A M (1994), Significant biogenesis of chlorinated aromatics by fungi in natural environments. *Appl. Environ. Microbiol.* **60**, 264–70.

DU M and AHN D U (2002), Effect of antioxidants on the quality of irradiated sausages prepared with turkey thigh meat. *Poultry Sci.* **81**, 1251–56.

EYER H, GAUCH R and BOSSET J O (1990), The fishy off-flavour of coffee cream. *Schweiz Milchwissenschaft und Forschung* **19**, 43–5.

FARES K, LANDY P, GUILARD R and VOILLEY A (1998), Physicochemical interactions between aroma compounds and milk proteins: effect of water and protein modification *J. Dairy Sci.* **81**, 82–91.

GASSENMEIER K (2002), A new pathway leading to cardboard off-notes in ice cream and how to protect flavour integrity. 10th Weurman Flavour research Symposium (25–28 June 2002, Beaune, F).

GIJS L, PIRAPREZ G, PERPÈTE P, SPINNLER E and COLLIN S (2000), Retention of sulfur flavours by food matrix and determination of sensorial data independent of the medium composition. *Food Chem.* **69**, 319–30.

HOENICKE K, BOERCHET O, GRÜNING K and SIMAT T J (2002), Untypical aging off-flavor in wine: synthesis of potential degradation compounds of indole-3-acetic acid and kinurenine and their evaluation as precursors of 2-aminoacetophenone. *J Agric. Food Chem.* **50**, 4303–9.

HORWOOD J F, LLOYD G T, RAMSHAW E H and STARK W (1981), An off-flavour associated with the use of sorbic acid during Feta cheese maturation. *Austr. J. Dairy Technol.* 38–40.

JACOBSEN C, HARTVIGSEN K, HOMSEN M K, HANSEN L F, LUND P, SKIBSTED L H, HOLMER G, ADLER NISSEN J and MEYER A S (2001), Lipid oxidation in fish oil enriched mayonnaise: calcium disodium ethylenediaminetetraacetate, but not gallic acid strongly inhibited oxidative deterioration. *J. Agric Food Chem.* **49**, 1009–19.

KHEADR E E, VUILLEMARD J C and EL-DEEB S A (2002), Acceleration of Cheddar cheese lipolysis by using liposome-entrapped lipases *J. Food Sci.* **67**, 485–92.

LEMIEUX L and SIMARD RE (1991), Bitter flavour in dairy products. I. A review of the factors likely to influence its development, mainly in cheese manufacture. *Lait* **71**, 599–636.

LOPEZ DA SILVA J A, CASTRO S M and DELGADILLO I (2002), Effect of gelatinization and strach-emulsifier interactions on aroma release from starch rich model system. *J. Agric. Food Chem.* **50**, 1976–84.

MATSUBARA H, GOTO K, MATSUMURA T, MOCHIDA K, IWAKI M, MOTOHIRO N and YAMASATO K (2002), *Alicyclobacillus acidiphilus* sp. nov., a novel thermo-acidophilic, w-alicyclic bacterium isolated from acidic beverages. *Internat. J. System. Evolution. Microbiol.* **52**, 1681–85.

MOLIMARD P, LESSCHAEVE I, BOUVIER I, VASSAL L, SCHLICH P, ISSANCHOU S and SPINNLER H E (1994), Amertume et fractions azotées de fromages à pâte molle de type camembert: rôle de l'association de *Penicillium camemberti* avec Geotrichum candidum. *Le Lait* **74**, 361–74.

O'CONNELL P F FOX (2001), Significance and applications of phenolic compounds in the production and quality of milk and dairy products: a review. *International Dairy J.* **11**, 103–20.

PALAMAND S R and GRIGSBY J H (1974), Stale flavors in beer. Identification of o-aminoacetophenone and ethylnicotinate in beer. *Brew. Dig.* **49**, 58–60.

REINECCIUS G (1991), Off-flavors in foods. *Crit rev. Food Sci. Nutrition.* **29**, 381–402.

REINERS J, NICKLAUS S and GUICHARD E (2000), Interaction between β-lactoglobulin and flavour compounds of different chemical classes. Impact of the protein on the odour perception of vanillin and eugenol. *Lait*, **80**, 347–60.

SCHWARZ P, STANLEY P and SOLBERG S (2002), Activity of lipase during mashing. *J. Am. Soc. Brew. Chem.* **60**, 107–9.

SHINOHARA T, KUBODERA S and YANAGIDA F (2000), Distribution of phenolic yeasts and production of phenolic off-flavors in wine fermentation, *J. Bioscience and Bioengineering* **90**, 90–7.

SPINNLER H E and GROSJEAN O et al. (1992), Effect of culture parameters on the production of styrene (vinyl benzene) and 1-octene-3-ol by *Penicillium caseicolum*. *J. of Dairy Research* **59**, 533–41.

SURIYAPHAN O, CADWALLADER K R and DRAKE M A (2001), Lecithin associated off-aromas in fermented milk. *J. Food Sci.* **66**, 517–23.

UMMADI M and WEIMER B C (2001), Tryptophan catabolism in *Brevibacterium linens* as a potential cheese flavor adjunct. *J. Dairy Sci.* **84**, 1773–82.

9
Taints from cleaning and disinfecting agents

C. Olieman, NIZO food research, The Netherlands

9.1 Introduction

The processing of foods requires regular cleaning of equipment in order to ensure that food products are safe for the consumer and of consistent quality. The cleaning process consists essentially of two stages:

- cleaning: removal of organic and inorganic deposits
- disinfection: sanitising the equipment to kill pathogenic and spoilage bacteria.

Depending on the nature and amount of the deposits, cleaning and disinfecting can be performed in one or two steps. After cleaning, the cleaning and disinfecting agents are usually removed.

Although the purpose of cleaning and disinfecting agents is to improve food safety and quality, such agents can, potentially, become a hazard themselves if residues are left, either because they are toxic or because they cause taints. It has been estimated that up to 30% of food taint complaints are associated with cleaning and disinfecting chemicals, producing taints variously described as 'soapy', 'antiseptic' or 'disinfectant' (Holah, 1995). They can enter food products accidentally, for example from poor rinsing or aerial transfer. However, they can also result from 'no rinse' disinfectants designed to be left on surfaces to provide more lasting protection against recontamination.

Legislation in Europe is divided on whether or not disinfectants can be left on surfaces without rinsing. The Meat Products Directive (95/68/EC) allows disinfectants to remain on surfaces 'when the directions for use of such substances render such rinsing unnecessary'. In contrast, the Egg

190   Taints and off-flavours in food

Products Directive (89/437/EC) and the Milk Products Directive (92/46/EEC) require that disinfectants be rinsed off fully with potable water. In general, 'no rinse' disinfectants need to be rigorously tested before use, and should not be used where there is direct contact with food which might give rise to taint. Indeed, it has been argued that low concentrations of disinfectants left on surfaces may present an inadequate biocidal challenge to microbial contamination and might even encourage the formation of resistant microbial populations on surfaces (Holah, 2000).

This chapter looks at the various cleaning and disinfectant agents used in food processing, how they can be tested for toxicity, their potential to cause taints, and the methods available for detecting their presence in rinse water or food products.

9.2   Cleaning and disinfecting agents

There exists a large variety of cleaning and disinfecting products on the market. However, the majority is based on one or more of the ingredients listed in Table 9.1. These are divided into cleaning and disinfecting agents. In practice, some ingredients combine both functions. As an example, although active chlorine-containing agents are mainly used for disinfection,

Table 9.1   Basic ingredients of cleaning and disinfecting agents

Ingredient	Function	Concentration
Acid (e.g. nitric, phosphoric acid)	Removal of inorganic deposits	
Alkaline (e.g. sodium hydroxide)	Removal of organic deposits (proteins, fat, carbohydrates)	
Sequestrants (e.g. EDTA)	Removal of inorganic deposits	
Quaternary ammonium compounds (e.g. didecyldimethylammonium chloride, alkyldimethyl-benzylammonium chloride)	Disinfecting, removal of fat	0.05–2%
Active chlorine (e.g. sodium hypochlorite, Chloramine T, sodium dichloroisocyanurate)	Disinfecting	0.015–0.03% (active chlorine)
Active iodine (iodophor)	Disinfecting	0.005–0.01% (active iodine)
Active oxygen (e.g. hydrogen peroxide with/without peracetic acid)	Disinfecting	0.03–0.5% (active oxygen)

Table 9.2 Cleaning and disinfecting agents: relative toxicity and risk of taint

Ingredient	Toxicity	Risk of taint
Acid	low	medium
Alkaline	low	medium
Sequestrants	low	low
Quaternary ammonium compounds	low	low
Active chlorine	medium	medium
Active iodine	medium	medium
Active oxygen	medium	medium

they also remove certain deposits. Similarly, alkaline solutions are used to remove organic deposits but also have a disinfecting action.

Most cleaning agents are composed of chemical ingredients which are not reactive. The cleaning is based on a physical interaction between, for example, surfactants and the deposited material which results in a solubilisation of the deposit. As a result, whenever residues of these ingredients come into contact with the food their impact on the product can vary from modest to negligible, depending on the concentration of the residual cleaning agent. The risk of toxicity or taint at normal levels of use is usually low. The relative safety and risk of taint from cleaning and disinfectant agents is summarised in Table 9.2.

The situation is different when disinfectants are used. Disinfectant agents themselves are more likely, if residues survive in sufficient concentrations, to present a potential safety risk and risk of taint. Some disinfectants, particularly when they are based on active chlorine, iodine or oxygen, present additional problems. These are reactive chemicals which can react with food components to form new components. Some of those new components can then cause off-flavours. As an example, chlorine and iodine-based disinfectants can react with food components to form chloro-phenols and iodo-phenols. These generate off-flavours and have a very low sensory threshold. Concentrations of a few parts per billion (ppb) produce serious off-flavours.

Active chlorine and iodine react particularly with methylketones in food to form, respectively, chloroform and iodoform. This so-called Haloform reaction is shown in Fig. 9.1 (Roberts *et al.*, 1967). Methylketones are present in low concentrations in most foods where they often contribute to the characteristic flavour of the product. The Haloform reaction is often rapid and is dependent on the pH of the food (Tiefel *et al.*, 1997). The formation of chloroform is also influenced by other components of the disinfecting agent such as quaternary ammonium and sequestering compounds. The sequestering agents nitrilotriacetic acid (NTA) and ethylenediaminetetraacetic acid (EDTA), in combination with hypochlorite, decrease the formation of chloroform. Quaternary ammonium compounds increase the

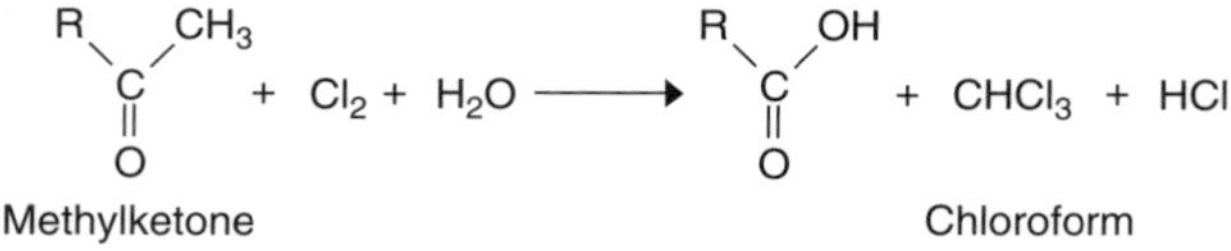

$$\underset{\text{Methylketone}}{\text{R–C(=O)–CH}_3} + Cl_2 + H_2O \longrightarrow \underset{\text{}}{\text{R–C(=O)–OH}} + CHCl_3 + HCl$$

Fig. 9.1 Haloform reaction.

formation of chloroform (Tiefel *et al.*, 1997). Chloroform is considered a potential carcinogen. In Germany, for example, the maximum allowable concentration of halogenated hydrocarbons in food products is limited to 0.1 mg/kg (LHmV, 1989).

9.3 Testing the safety of cleaning and disinfecting agents

As far as demonstrating the non-toxicity of cleaning agents is concerned, legislation in Europe has in the past varied between member states. A recognised industry guideline for disinfectants is a minimum acute oral toxicity (with rats) of 2000 mg/kg bodyweight. The implementation of the Biocidal Products Directive (98/9/EC) in 2000 has introduced greater consistency between European member states.

Annexe I of the Directive lists all the permitted biocidally-active substances known to the European market. Annexes II and III list the data and tests required for a biocidal product to be authorised for inclusion in Annex I. These include:

- formulation
- data on physical and chemical properties
- intended uses
- classification and labelling
- effectiveness against target organisms
- effect of residues on food
- toxicological profile and health-related studies
- ecotoxicological profiling.

The Directive also establishes a product authorisation scheme for new products not already included in Annexe I. The Directive states that 'Member states shall prescribe that a biocidal product shall not be placed on the market and used in their territory unless it has been authorised in accordance with the Directive.' Authorisation involves submitting data on formulation, physical and chemical properties, and proving the suitability of a biocidal product according to the criteria set out in Annexes II and III. Food processors should ensure that the biocidal products they use have been registered according to the terms of the Directive.

9.4 Testing cleaning and disinfecting agents for their capacity to cause taints

There are a number of ways of testing whether disinfectant residues may cause taints. The Campden and Chorleywood Food Research Association has developed two taint tests in which foodstuffs exposed to disinfectant residues are compared with control samples using a standard triangular taste test (Anon., 1983a). The results of these tests are statistically assessed to isolate any flavour difference and to describe the nature of the taint (see Chapter 2).

To assess the potential aerial transfer of a taint from a disinfectant to a foodstuff, a modification of a standard packaging materials odour transfer test is used (Anon., 1964). This involves using samples of four types of food with differing levels of susceptibility to aerial taint:

- high moisture (e.g. fruit)
- low moisture (e.g. biscuits)
- high fat (e.g. cream)
- high protein (e.g. chicken).

Samples are suspended over a disinfectant solution using distilled water for 24 hours before being assessed by the taint panel.

A modification of the standard food container transfer test is used to test surface transfer (Anon., 1983b). This involves either:

- spraying disinfectants onto two sheets of stainless steel followed by rinsing; or
- spraying disinfectants onto the stainless steel sheets and draining the disinfectant residue off to simulate a 'no rinse' application.

Food samples are sandwiched between the two stainless steel sheets and left for 24 hours before being compared by the taint panel to control sheets rinsed in distilled water only.

9.5 Detecting cleaning and disinfecting agents in rinse water

Detecting cleaning and disinfecting agents in rinse water requires particular techniques. As cleaning and disinfecting agents lack chromophores that adsorb visible light, a simple visual inspection or *in situ* spectrophotometric measurement of the rinse water is not effective. In the case of cleaning with alkali- or acid-based cleaning agents, pH measurement of the rinse water provides a simple indication of how successful rinsing has been.

Since the majority of cleaning agents have ionic properties, conductometry techniques can be used to monitor rinse water for cleaning agents. The rinsing of the system is successful if the conductivity of the rinse water effluent is below a certain threshold. Conductometric probes are of rugged design and can be easily integrated into food processing operations to check rinse water for the presence of residual cleaning agents. In some cases manufacturers add additional salt to the cleaning agent in order to make monitoring by conductometry more sensitive.

Conductometry techniques are not suitable for measuring traces of active chlorine, iodine and oxygen-containing detergents. Off-line photometric methods are appropriate for the determination of active chlorine, active oxygen, quaternary ammonium compounds, halogenated acids and formaldehyde, and are used, for example, in the brewing industry (Treetzen et al., 1989).

9.6 Detecting cleaning and disinfecting agents in food

Detection by conductometry of residual cleaning agents in a food product is more difficult than in rinse water, because the conductivity of the product is generally higher than that of the water used for rinsing. An increase in conductivity, which would clearly indicate the presence of residual cleaning agents in rinse water, is difficult to detect against the background of natural variation in conductivity in the food product. Another problem is fouling of the conductivity probe.

Detecting residues of disinfecting agents in food products is even more difficult. Detection of traces of strong alkaline or acid agents is often limited by the presence of the same cations or anions in the sample food matrix. A similar problem affects some active chlorine compounds. Hypochlorite decomposes to chloride which is often present in the matrix. Hydrogen peroxide decomposes to water and is therefore also not easily detectable. Peracetic acid decomposes to acetic acid which is often present in the matrix, limiting the detection of this component.

The situation is easier for quaternary ammonium compounds, Chloramine T, dichloroisocyanurate and iodophors. These agents leave behind specific residues, which can be detected by methods based on high performance liquid chromatography (HPLC). Para-toluenesulfonamide is the decomposition product of Chloramine T. It has been determined by reversed-phase HPLC in fish fillets (Meinertz et al., 1999, 2001) and by a combination of continuous flow and liquid chromatography in ice cream (Beljaars et al., 1994). Iodophors decompose to iodide which is present as a trace element in most food products. However, measurement is possible after anion-exchange chromatography in combination with suppressed conductivity detection, followed by pre-column derivatisation and reversed phase HPLC (Verma et al., 1992). Measurement has also been undertaken for milk by reversed-phase ion-pair separation in combination with electrochemical detection at a silver electrode (Sertl et al., 1993).

9.7 Measurement of active chlorine residues via chloroform

Of all possible components which can be formed by the action of active chlorine in a food product, chloroform is probably the most abundant and it can be relatively easily measured. Chloroform is a lipophilic substance. In whole milk (fat content of ca. 3.5%), for example, chloroform resides mainly in the fat phase. It is a volatile component and forms with water an azeotrope with a boiling point of 56.3°C and a composition of 97% of chloroform in the vapour phase (Weast, 1980). This means that on heating a whole milk sample in closed vial to ca. 56°C an appreciable amount of the chloroform is present in the vapour phase, making a simple sampling procedure possible (Fig. 9.2). Gas chromatography in combination with a halogen-specific detector enables a sensitive measurement of chloroform and other chlorinated hydrocarbons down to ca. $0.1\,\mu g/l$.

The formation of chloroform and other chlorinated hydrocarbons has been studied for milk in an experimental set-up simulating practical CIP conditions (Linderer et al., 1994). The study used different concentrations of disinfectant with and without rinsing with water. Although the study did not specify which type of active chlorine disinfectant was used, chloroform concentrations of 1.5 to $5.4\,\mu g/l$ were found for whole milk, 0.7 to $2.2\,\mu g/l$ for skim milk and 13 to $40\,\mu g/l$ for cream. Using this method, an overview of concentrations of chloroform in commercial butter and milk samples is given by Westermair (Westermair, 1998).

Recent developments in the detection of chloroform include mass spectrometry. Mass spectrometry has evolved rapidly as a technique during the last 20 years towards simple-to-use laboratory systems and rugged process gas analysers. Direct head-space measurement of chloroform now is possible, for example, with the Food-sense, a further development of the Airsense 2000 manufactured by V&F (Absam, Austria). This apparatus uses a patented soft chemical ionisation technique which controls the contact of the moist sample gas with the delicate filament. The ions are formed in a charge-exchange chamber and subsequently separated on their mass over charge ratio (m/z) by a quadropole. This set-up results in a highly stable and sensitive mass spectrometer, showing detection limits to low ppb-values for all kind of volatile or gaseous substances. At NIZO food research the measurement of chloroform by direct head-space sampling has been successfully evaluated using the Food-sense.

Chlorine atoms exist as two isotopes having masses of 35 (75.5%) and 37 (24.5%). On ionisation chloroform loses a chlorine atom. The residual

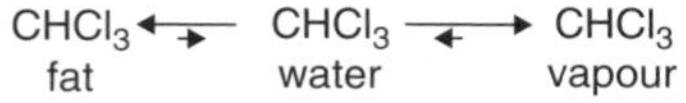

$$CHCl_3 \text{ (fat)} \leftrightarrow CHCl_3 \text{ (water)} \leftrightarrow CHCl_3 \text{ (vapour)}$$

Fig. 9.2 Equilibrium of chloroform between the fat, water and vapour phases of milk. On increasing temperature the equilibrium shifts to the right.

positive ion $CHCl_2^+$ shows up at masses 83, 85 and 87 in decreasing intensity due to the isotopic composition of the chlorine. In principle measurement at mass 83, the most abundant signal, would suffice. However, when dealing with complex samples of natural origin, it is possible that another component might form an ion or ion-fragment having the same mass as the one monitored for chloroform. If chloroform is monitored at more than one mass, the likelihood of a false positive result diminishes rapidly. The Food-sense uses mercury vapour (the system is completely closed, using special filters in the exhaust of the vacuum pump) or xenon gas as ionisation gases, having ionisation energies of 10.4 eV and 12.1 eV, respectively. Most organic substances can be ionised using mercury, whereas xenon often induces more fragmentation. The instrument can switch within a second between ionisation with mercury or xenon. Both can be used to ionise chloroform, enabling the measurement at masses 83 and 85 using mercury and 85 and 87 using xenon (measurement at mass 83 using xenon is hampered by residual krypton in the xenon). Chloroform is judged to be present if all results are positive. In this way false positive results are unlikely.

The system is calibrated by adding known amounts of chloroform to the type of product to be measured (e.g. whole or partially skimmed milk). The

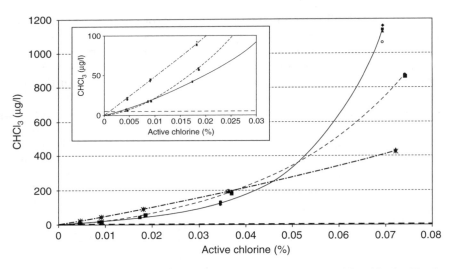

Fig. 9.3 Head-space measurement of chloroform in whole milk with the Food-sense. Chloroform formation was induced by the addition of sodium hypochlorite (short-long dashed line), dichloroisocyanurate (small dash line), and Chloramine T (continuous line) to whole milk. Chloroform was measured at mass 83 using Hg (♦), mass 85 using Hg (■), mass 85 using Xe (▲) and mass 87 using Xe (○). Horizontal dashed line indicates detection limit (5 µg/l). The insert shows in greater detail the graph at low active chlorine and chloroform levels.

result of the addition of sodium hypochlorite, Chloramine T (Halamid, N-chloro-p-toluenesulfonamide sodium salt) and sodium dichloroisocyanurate to whole milk is shown in Fig. 9.3. The horizontal dashed line shows the detection limit (5 μg/l) of the system. Chloroform is detected at 5 μg/l when ca. 0.004% of active chlorine is present. The similar results obtained for the different isotopes and ionisation methods show that no interfering components have been formed. It is notable that at high concentrations of active chlorine, Chloramine T and dichloroisocyanurate generate considerably more chloroform than hypochlorite (Olieman et al., 2002). Since measurement takes less than one minute per sample, it provides an effective means of measuring large numbers of samples of raw milk quickly.

9.8 Future trends

Active chlorine disinfectants are being increasingly replaced by hydrogen-peroxyde with or without peracetic acid, because these disinfectants are less corrosive for stainless steel. Contact of these disinfectants with the food product could lead to taints and off-flavours. In principle it should be possible to detect in-line off-flavours or deviations in the normal pattern of volatiles by sensitive mass spectrometry or by using fast gas chromatography in combination with mass spectrometric detection. These techniques are now becoming available as on-line process control instruments.

9.9 References

ANON. (1964), Assessment of odour from packaging material used for foodstuffs. British Standard 3744. British Standards Institution, London.
ANON. (1983a), Sensory analysis of food. Part 5: Triangular test. ISO 4120. International Standards Organisation.
ANON. (1983b), Testing of container materials and containers for food products. DIN 10955. Deutsches Institut für Normung e.V., Berlin.
BELJAARS P R et al. (1994), 'Determination of p-toluenesulfonamide in ice cream by combination of continuous flow and liquid chromatography: summary of collaborative study', *J. of the AOAC International*, **77**, 672–4.
HOLAH J (1995), 'Special needs for disinfectants in food-handling establishments', *Revue Scientifique Office Internationale des Epizooties*, **14**, 95–104.
HOLAH J (2000), 'Cleaning and disinfection', in Stringer M and Dennis C (eds), *Chilled foods: a comprehensive guide (Second edition)*, Woodhead Publishing Limited, Cambridge.
LHmV (1989), 'Grenzwerte in der Lösungsmittel-Höchstmengenverordnung (LHmV) für Lebensmittel'.
LINDERER M et al. (1994), 'Untersuchungen zur Problematik von Rückstanden in Milch nach Reinigungs- und Desinfectionmaßnamen' [Investigations of the problem of residuals in milk after cleaning and disinfecting], *Deutsche Milchwirtschaft*, **17**, 778–82.

MEINERTZ J R et al. (1999), Liquid chromatographic determination of para-toluenesulfonamide in edible fillet tissues from three species of fish, *J. of the AOAC International*, **82**, 1064–70.

MEINERTZ J R et al. (2001), Performance of a proposed determinative method for p-TSA in rainbow trout fillet tissue and bridging the proposed method with a method for total chloramine-T residues in rainbow trout fillet tissue, *J. of the AOAC International*, **84**, 1332–6.

OLIEMAN C et al. (2002).

ROBERTS J D et al. (1967), *Modern Organic Chemistry*, New York, W.A. Benjamin.

SERTL D et al. (1993), Liquid chromatographic method for determination of iodine in milk – collaborative study, *J. of AOAC International*, **76**, 711–19.

TIEFEL P et al. (1997), Model tests for the formation of chloroform by chlorine containing cleaning and disinfecting products, *Milchwissenschaft*, **52** (12), 686–90.

TREETZEN U et al. (1989), Bestimmung von Desinfektionsmitteln in Spülwasser [Determination of disinfectants in wash water], *Monatsschrift für Brauwissenschaft*, **42**, 211–18.

VERMA K K et al. (1992), Determination of iodide by high-performance liquid chromatography after pre-column derivatisation, *Analytical Chemistry*, **64**, 1484–9.

WEAST R C (1980), *CRC Handbook of Chemistry and Physics*, Boca Raton, CRC Press, Inc.

WESTERMAIR T (1998), Rückstände an Reinigung- und Desinfektionsmitteln in Milch und Milchprodukten [Residues in milk and milk products of compounds used for cleaning and disinfection], *Deutsche Milchwirtschaft*, **49**, 703–6.

Index

Achromobacter 118, 122–3, 124–5
Acinetobacter 117–18
actinomycetes 129–30, 136
active packaging 108
additives
   bacterial interactions with 184–5
   in food packaging 73–4
adhesives 73
aerobic bacteria 117–28
affective tests 12–13, 14, 16–20
Alcaligenes 118
Alicyclobacillus 118–19, 183
Alteromonas 118
aluminium foil 68–9
ammonium compounds 194
anaerobic bacteria 117, 128–9
   facultative 125–8
analytical tests 12–14
androstenone 34
antioxidants 144–5
appearance of food 1, 7
application-specific electronic noses
      (ASEN) 48–53
aroma scalping 67
aromas and odours 1, 2, 8, 65–6
   DIN standards 103
   food staling 170–3
   sensory properties 170
   thresholds 70, 168
Aspergillus 131–2
atomic emission detectors 44
autoxidation 143–6, 153–6, 169–70

Bacillus 125
bacteria
   Achromobacter 118, 122–3, 124–5
   Acinetobacter 117–18
   additive interactions 184–5
   aerobic bacteria 117–28
   Alcaligenes 118
   Alicyclobacillus 118–19, 183
   Alteromonas 118
   anaerobic bacteria 117, 128–9
      facultative 125–8
   Bacillus 125
   Brevibacterium 183
   Brochothrix 117, 125–6
   Clostridium 128–9
   food matrix interactions 182–3
   Lactobacillus 117
   Lactococcus 183
   Moraxella 119
   off-flavours in particular foods
      114–16
   Photobacterium 117
   Pseudomonas 117, 119–25
   Rahnella 127–8
   Serratia 126–7
   Shewanella 117, 118, 124–5
   Yersinia 128
beef 114, 129
beer 44, 53, 183
bipolar scales 17
bitterness 9, 182
boar taint 129

boiled sweets bags 90–7
bread bags 89–90
Brevibacterium 183
Brochothrix 117, 125–6
bromoanisoles 79
butyric acid 183

cake mixes 33
capsaicin 8
carbohydrates 140
cardboard packaging 69
category scales 17, 18
catfish 35, 36
cationic inks 107
cellulose films 70
cereals 116
cheese 35, 130, 183–4
   packaging 97–101
chemical ionisation 47, 80
chemical irritants 8, 10
chicken 115, 153
chilli 8
chlorine 182
chloroform measurement 195–7
chlorophenol contamination 6
chocolate 78
clarifying agents 73–4
cleaning and disinfecting agents
      189–97
   chloroform measurement 195–7
   detecting
      in food 194–5
      in rinse water 193–4
   ingredients in 190–2
   and legislation 189–90, 192
   no rinse disinfectants 189–90
   safety tests 192
   testing 192–3
closed-loop stripping apparatus
      (CLSA) 40
Clostridium 128–9
co-extrusions 108
codes of practice 26–7, 101–4
   see also legislation
coffee 35, 162, 172
colour of food 1, 7
conductometric probes 194
consumer acceptability testing 19–20
contracting out testing work 24
cork taint 29, 34, 41, 43

data analysis and presentation 20, 22
detection thresholds 10
diagnostic taint testing 23–4

difference tests 14–16
digital printing 107–8
DIN standards 103
directives *see* legislation
discrimination tests 14–16
disinfecting agents *see* cleaning and
      disinfecting agents
distillation 34–7, 85–7
duo-trio tests 15
dynamic headspace sampling 38–40,
      81–4, 100

eggs 115
electron capture detectors (ECD) 32,
      42, 44
electronic noses 29, 48–54, 104–6
ethics 28
Eurotium 132–3

facultative anaerobic bacteria 125–8
fat content of foods 143
fingerprint mass spectra systems (FMS)
      51–3
fish 35, 36, 40, 42, 116, 156–7
flame photometric detectors (FPD) 44
flavour 1, 8, 162–3
   compound volatility 176–81
   retention 180
   thresholds 168
   Yaylayan classification 166–7
flow wraps 74
food matrices 176–86
   bacterial interactions 182–3
      with additives 184–5
   flavour compound volatility 176–80
   flavour retention 181
   prevention 185–6
   reactions between components
      182–3
food staling 170–3
free-choice profiling 19
fruit juices 42, 171
fruit and vegetables 116
fungi 130–5
   Aspergillus 131–2
   Eurotium 132–3
   growth of 130–1
   Penicillium 133–5, 182–3

gas chromatography with selective
      detectors 43–5
gas chromatography-mass spectrometry
      45–7, 80–1

Index

gas chromatography-olfactometry 42–3, 80–1
Geotrichum 182
glass packaging 67
graphic scales 17
guaiacol 34, 35

Haloform reaction 191
headspace extraction 37–40, 81–5
hedonic scales 17
hedonic/affective tests 12–13, 14, 16–20
hemoproteins 155–6
high resolution gas chromatography (HRGC) 32
high-performance liquid chromatography (HPLC) 33, 47
human senses 1–2, 6–9
hydroperoxides 145–6, 147–8
hypothesis testing 22

ice cream 185
in store printing 108
infrared scanning detectors 80
instrumental analysis 29, 31–54
   atomic emission detectors 44
   closed-loop stripping 40
   dynamic headspace sampling 38–40, 81–4, 100
   electronic noses 29, 48–54, 104–6
   fingerprint mass spectra 51–3
   flame photometric detectors 44
   gas chromatography with selective detectors 43–5
   gas chromatography-mass spectrometry 45–7, 80–1
   gas chromatography-olfactometry 42–3, 80–1
   headspace extraction 37–40, 81–5
   high-performance liquid chromatography 33, 47
   liquid-based extraction 32–7, 87–8
   and packaging 80–1
   solid phase microextraction 40–2, 81, 85
   solvent extraction 33, 36–7
   static headspace sampling 38, 85
   steam distillation 34–7, 85–7
   stir-bar sorptive extraction 47–8
   supercritical fluid extraction 33–4
internet information sources 109
ion traps 80
iron atoms 151
irradiation 181–2

ketonic rancidity 151–2
kinesthesis 8
Kurdurna-Danish evaporation apparatus 86–7

Lactobacillus 117
Lactococcus 183
lamb 114
laminates 66–7
legislation 101–2, 189–90, 192
Likens–Nickerson extractor 36–7, 81, 85–6
lipids 140–1, 169, 181
   see also rancidity
lipoxygenases 149–51
liquid-based extraction techniques 32–7, 87–8
litho printing 72–3
litigation 24

Maillard reaction 162–73
malodours 65
margarine 42
Marks & Spencer 27
Mason jars 44
mass spectrometers 32, 42, 43, 45–7, 80–1, 94, 195
mastication 8–9
measuring human responses 9
meat 153–6
   beef 114, 129
   chicken 115, 153
   hemoproteins in 155–6
   lamb 114
   pig fat and meat 33, 34, 47, 114
melons 35
metal packaging 69
metal-catalysed lipid oxidation 151–2
microbiological spoilage 2, 112–36
   actinomycetes 129–30, 136
   fungi 130–5
   preventing 157–8
   Streptomyces griseus 130
   see also bacteria
microwave-assisted steam distillation 35–6
milk 42, 47, 53, 114, 148, 171
   UHT treatment 44
moisture control systems (MCS) 84
Moraxella 119
moulds 151

odours *see* aromas and odours
off-flavours, definition 2, 5–6

off-odours 65
Oxford Chemicals Ltd 101
oxidation 142–3, 156–7

packaging 2, 64–109
  active packaging 108
  additives in 73–4
  alterations to 74
  boiled sweets bags 90–7
  bread bags 89–90
  cheese 97–101
  chemicals responsible for taint 75–8
  codes of practice/standards 27, 101–4
  and electronic noses 104–6
  food affected by taint 78–9
  Framework Directive 101–2
  information sources 109
  instrumental analysis 80–1
  prevention 101–4
  printing inks 71–3, 74, 107–8
  PVC odours 100–1
  sample preparation 81–8
  sampling strategy 88
  smart packaging 108
  sources of taints 70–5
  taint transfer tests 25–6
  tracing causes of taint 106–7
  types of food packaging 66–70
paired comparison tests 14–15
panel tests 13–14
  see also sensory tests
paper laminates 66, 68–9
paperboard packaging 67–8
peanuts 172
Penicillium 133–5, 182–3
phenol 70, 79
phenolic derivatives 181
photo-oxidation 146–9
Photobacterium 117
pig fat and meat 33, 34, 47, 114
plastic laminates 66–7, 68–9, 71
polyethylene 67
polymers 67
polyolefins 67
polyunsaturated fatty acids 156
pork 33, 34, 47, 114
preservative agents 182
preventive testing 24–6
printing inks 71–3, 74, 107–8
Procrustes analysis 19
profile tests 18–19, 20
progressive profiling 19
promotional printed items 74
proteins 140, 181

Pseudomonas 117, 119–25
  Pse. fluorescens 120–1
  Pse. fragi 121–2
  Pse. graveolens 124
  Pse. perolens 122–3
  Pse. putida 123
  Pse. putrefaciens 124–5
  Pse. taetrolens 124
purge and trap equipment 84, 96–7
PVC odours 100–1

qualitative tests *see* hedonic/affective tests
quantitative descriptive analysis (QDA) 18–19

R-index test 16
Rahnella 127–8
rancidity 140–58
  autoxidation 143–6, 153–6, 169–70
  ketonic rancidity 151–2
  lipoxygenases 149–51
  metal-catalysed lipid oxidation 151–2
  oxidation 142–3, 156–7
  photo-oxidation 146–9
  volatile lipid molecules 152–3
recognition thresholds 10
regenerated cellulose films 70
residual solvent analysis 103–4
riboflavin 147, 148
roasted foods 171–2
Robinson test 103

sample preparation 81–8
sampling strategy 88
scaling procedures 17–18, 102, 103
scratch-off cards 74
screen printing 74
selected ion monitoring (SIM) analysis 45–6, 85
sensor array technology (SAT) 53
sensory tests 5–29, 102–4
  analytical tests 12–14
  applications of 23–6
  attributes of interest 18–19
  consumer acceptability testing 19–20
  data analysis and presentation 20, 22
  diagnostic taint testing 23–4
  difference/discrimination tests 14–16
  duo-trio tests 15
  environment 11
  ethics in 28

hedonic/affective tests 12–13, 14, 16–20
instrumental analysis 29
measuring human responses 9
objectives of 10–1
paired comparison tests 14–15
panels of assessors 13–14
preventive testing 24–6
profile tests 18–19, 20
R-index test 16
scaling procedures 17–18, 102, 103
selection and interpretation of tests 20–2
standards in 26–7
taint transfer tests 24–7
test subjects 11–14
3-AFC (alternative forced choice) tests 15–16
threshold measurements 10, 70–1, 106–7
time-intensity tests 19
triangle tests 15, 105
Serratia 126–7
Shewanella 117, 118, 124–5
sizing agents 68
skatole 33, 34
slip agents 73
smart packaging 108
solid phase microextraction 40–2, 81, 85
solvent extraction 33, 36–7
somesthesis 8
sorbic acid 38, 185
staling 170–3
standards 26–7, 101–4
  see also legislation
static headspace sampling 38, 85
steam distillation 34–7, 85–7
stir-bar sorptive extraction 47–8

storage conditions 177
strawberry juice 42
Strecker degradation 165
Streptomyces griseus 130
supercritical fluid extraction 33–4

taint, definition 2, 5–6, 65
taint transfer tests 24–7
taste 1, 7–8
  thresholds 71
Tenax 39, 41, 82, 84
test subjects 11–14
texture 1, 8
thermodynamics 177
3-AFC (alternative forced choice) tests 15–16
thresholds 10, 70–1, 106–7
  aromas and odours 70, 168
  flavour 168
  taste 71
time-intensity tests 19
touch stimuli 8
triangle tests 15, 105

unipolar scales 17

vacuum distillation 35, 37
varnishes 71–3
visual senses 7
volatile lipid molecules 152–3
volatile sensors 29

water taints 40
wine 35, 184
  cork taint 29, 34, 41, 43
wood preservatives 70, 79

yeasts 184
Yersinia 128

Account
403450 2003 0996